C000026495

IMPERIAL GAZETTEER OF INDIA

PROVINCIAL SERIES

BALUCHISTĀN

*Property of
Tony John,*

Mr Anthony John
8 Robingoodfellows Lan
MARCH
PE15 8HS

Elibron Classics
www.elibron.com

This Book is the Property of:

Tony John
8 Robingoodfellows Lane
March
Cambridgeshire
PE15 8HS

Elibron Classics series.

© 2005 Adamant Media Corporation.

ISBN 1-4212-4825-5 (paperback)
ISBN 1-4212-4824-7 (hardcover)

This Elibron Classics Replica Edition is an unabridged facsimile
of the edition published in 1908,
Calcutta.

Elibron and Elibron Classics are trademarks of
Adamant Media Corporation. All rights reserved.

This book is an accurate reproduction of the original. Any marks, names, colophons, imprints, logos or other symbols or identifiers that appear on or in this book, except for those of Adamant Media Corporation and BookSurge, LLC, are used only for historical reference and accuracy and are not meant to designate origin or imply any sponsorship by or license from any third party.

Mr Anthony John
8 Robingoodfellows Lane
MARCH
PE15 8HS

IMPERIAL GAZETTEER OF INDIA

PROVINCIAL SERIES

BALUCHISTĀN

SUPERINTENDENT OF GOVERNMENT PRINTING
CALCUTTA

1908

Price Rs. 2, or 3s.]

Mr Anthony John
8 Robingoodfellows Lane
MARCH
PE15 8HS

PREFACE

THE articles contained in this volume were compiled by Mr. R. Hughes-Buller, I.C.S. The collection of material entailed special difficulties, owing to the large area to be covered and the less highly organized staff of officials as compared with other parts of India. Mr. Hughes-Buller received valuable assistance from Major C. Archer, Indian Army, Revenue Commissioner in Baluchistān; and he also wishes to acknowledge the help given him by Rai Sāhib Dīwān Jamiat Rai in collating information about what has hitherto been an unknown country. The sections on Geology were contributed by Mr. E. Vredenburg of the Geological Survey of India, and those on Botany by Lieut.-Col. D. Prain, Director of the Botanical Survey of India (now Director of the Royal Botanic Gardens, Kew).

TABLE OF CONTENTS

PROVINCIAL GAZETTEERS
OF INDIA

BALUCHISTĀN

Baluchistān (more correctly Balochistān).—An oblong stretch of country occupying the extreme western corner of the Indian Empire, and situated between 24° 54′ and 32° 4′ N. and 60° 56′ and 70° 15′ E.

It is divided into three main divisions: (1) British Baluchistān, with an area of 9,476 square miles, consisting of tracts assigned to the British Government by treaty in 1879; (2) Agency Territories, with an area of 44,345 square miles, composed of tracts which have, from time to time, been acquired by lease or otherwise brought under control and been placed directly under British officers; and (3) the Native States of Kalāt and Las Bela, with an area of 78,034 square miles.

Baluchistān is bounded on the south by the Arabian Sea; Boundaries. on the north by Afghānistān and the North-West Frontier Province; on the west by Persia; and on the east by Sind, the Punjab, and a part of the Frontier Province. The western boundary from Gwetter Bay to Kuhak was settled by Colonel Goldsmid in 1871. A line from Kuhak to Koh-i-Malik-Siāh was defined by an Anglo-Persian Boundary Commission in 1896, and the southern portion of it was demarcated by pillars to the bank of the Talab river. There has been no demarcation north of that point, but the line thence to Koh-i-Malik-Siāh is governed by the agreement of 1896, and a supplementary agreement concluded in May, 1905. The Baloch-Afghān Boundary Commission delimited the northern frontier between 1894 and 1896. The boundary dividing Baluchistān from the Frontier Province on the one hand and the Punjab on the other has been defined at various times since the establishment of the Agency. That between Sind and Baluchistān was settled in 1854 and demarcated in 1862.

The Province covers a total area of 131,855 square miles, Dimensions. including the Native States of Kalāt and Las Bela, and is the largest of the Agencies administered under the Foreign

Department. Its area exceeds that of the whole of the British Isles. The country, which is almost wholly mountainous, lies on the great belt of ranges connecting the Safed Koh with the hill system of Southern Persia. It thus forms a watershed, the drainage of which enters the Indus on the east and the Arabian Sea on the south, while on the north and west it makes its way to those inland lakes or *hāmūns* which form so general a feature of Central Asia.

Origin of name.

The name of the country is derived from the Baloch, whose migratory hordes gradually extended eastwards from Southern Persia in and after the seventh century, until they eventually took up a position in Kachhi about the fifteenth century. The Baloch are not, however, the most numerous people in the Province, being exceeded in numbers by both Brāhuis and Afghāns.

Natural divisions.

The characteristic divisions of the country are four in number: upper highlands, lower highlands, plains, and deserts. The upper highlands, locally known as Khorāsān, occupy the central and east-central portion of the country, extending between 28° and 31° N. Here the mountains reach an elevation of nearly 12,000 feet, while the valleys lie about 5,000 feet above sea-level. The lower highlands include the slopes of the Sulaimān range on the east, the Pab and Kīrthar ranges on the south, and the ranges of Makrān, Khārān, and Chāgai on the west. The elevation of the valleys in this tract varies from 250 feet above sea-level upwards. The plains of Baluchistān include the peculiar strips of country known as Kachhi and Las Bela, and the valley of the Dasht river. They may be described as flat triangular inlets of generally similar formation, running up into the mountains. Their population differs markedly from that of the highlands. The deserts are situated in the north-western part of the Province. They consist of open level plains covered with black gravel, or of broad expanses of deep sand-hills which sometimes assume the proportions of formidable sand-mountains.

Mountain system.

The general configuration of the mountains resembles the letter S. On the east the SULAIMĀN range stretches upwards in gradually ascending steps to the Takht-i-Sulaimān. The mountains then curve round in a westerly direction on the northern side of the Zhob river along the TOBA-KĀKAR hills till the CENTRAL BRĀHUI range is reached. Near Quetta the direction becomes north and south, but from about the 66th degree of longitude the general trend is again in a westerly direction through Makrān and Khārān. To the south of the Central Brāhui range the KĪRTHAR and PAB ranges occupy the south-east corner of the Agency. On the west four parallel ranges occur, the southernmost being known as the MAKRĀN COAST range, the next as the CENTRAL MAKRĀN range, north of which again lies the SIĀHĀN range. Above these are situated the RĀS KOH, skirting Khārān, and the CHĀGAI hills. The mountains are, as a rule, composed of bare rocky limestone or conglomerate, and, except in the upper highlands, seldom have much vegetation. In southern Makrān the hills are distinguished by the absence of stones; and the white clay of which they consist has been worn by the lapse of ages into most fantastic shapes. A range seldom bears a distinctive title, but every peak is known by a separate name to the inhabitants.

No rivers are to be found carrying a large and permanent Rivers. flow of water. For the greater part of the year the beds contain merely a shallow stream, which frequently disappears in the pebbly bottom. Wherever practicable, this supply is taken off for irrigation purposes. After heavy rains the rivers become raging torrents; and woe to the man who happens to meet a flood in one of those weird gorges of stupendous depth, running at right angles to the general strike of the hills, which form so remarkable a feature of this region. Sometimes these defiles are so narrow that both sides can be touched at one time with the hands, and the walls rise many hundred feet perpendicularly upwards. The largest river in the country is the HINGOL or Gidar Dhor. The north-eastern part of the Province is drained by the ZHOB river on the east and the PISHĪN LORA on the west. Farther south the NĀRI receives the drainage of the Loralai and Sibi Districts and passes through Kachhi. The rivers draining the Jhalawān country are the MŪLA, the HAB, and the PORĀLI. In Makrān the DASHT river carries off the drainage to the south, and the RAKHSHĀN, which joins the Māshkel river, to the north.

The traveller who has left the plains of India and entered Scenery. the passes of Baluchistān finds himself among surroundings which are essentially un-Indian. The general outlook resembles that of the Irānian plateau, and, taken as a whole, it is un-attractive, though its peculiarities are not without a certain charm. Rugged, barren, sunburnt mountains, rent by huge chasms and gorges, alternate with arid deserts and stony plains the prevailing colour of which is a monotonous drab. But this is redeemed in places by level valleys of considerable size, in which irrigation enables much cultivation to be carried on and rich crops of all kinds to be raised. The flatness of the valleys, due to the scanty rainfall, distinguishes Baluchistān from the Eastern Himālayas.

Within the mountains lie narrow glens whose rippling water-courses are fringed in early summer by the brilliant green of carefully terraced fields. Rows of willows, with interlacing festoons of vines, border the clear water, while groups of ruddy children and comely Italian-faced women in indigo-blue or scarlet shifts and cotton shawls complete a peaceful picture of beauty and fertility. Few places are more beautiful than Quetta on a bright frosty morning when all the lofty peaks are capped with glistening snow, while the date-groves, which encircle the thriving settlements of Makrān, are full of pictur-esque attraction. The frowning rifts and gorges in the upper plateau make a fierce contrast to the smile of the valleys. From the loftier mountain peaks magnificent views are obtainable.

Lakes. No lakes of importance occur. The HĀMŪN-I-MĀSHKEL and HĀMŪN-I-LORA can hardly be described as such, for they only fill after heavy floods. The same may be said of the *kaps* of Parom and Kolwa in Makrān. The SIRANDA in Las Bela is a land-locked lagoon. Astālu or Haptālār, lying off the Makrān coast, is the only island, unless the bare rock of Churna off Rās Muāri be reckoned as such.

The coast. The Province has a coast-line of about 472 miles. The dis-tance in a direct line, however, from Karāchi to Gwetter Bay is only 335 miles. Owing to small rainfall, the salt nature of the soil, and the physical conformation of the country, the shore is almost entirely desert, presenting a succession of arid clay plains impregnated with saline matter and intersected by watercourses. From these plains rise precipitous table hills. The coast-line is deeply indented, but its most characteristic feature is the repeated occurrence of promontories and penin-sulas of white clay cliffs, table-topped in form. The inter-mediate tract is low, and in some places has extensive salt-water swamps behind it. The chief ports on the coast are Sonmiāni or Miāni, Pasni, and Gwādar. They are much exposed, and, owing to the shoaling of the water, no large ships can approach nearer than two or three miles.

Geology[1]. For geological purposes Baluchistān is conveniently divided into three regions :—

(1) An outer series of ranges, forming a succession of synclines and anticlines comparable in structure to the

[1] From material supplied by Mr. E. Vredenburg, Deputy-Superintendent, Geological Survey of India.

typical Jura mountains of Europe. In this region we
have two subdivisions : (*a*) the semicircular area of Sewistān[1], and (*b*) the ranges of Kalāt, Sind, and Makrān,
which continue into Southern Persia as far west as the
Straits of Ormuz. Further curved ranges of the same
type extend as far as Kurdistān.

(2) A region of more intense disturbance exhibiting the
typical Himālayan structure, whose southern or south-
eastern limit is a great overthrust forming the western
continuation of the Great Boundary Fault of the Himā-
laya[2]. It includes the ranges north of the Zhob and
Pishīn valleys. The Sewistān semicircular area stands
very much in the same relation to these ranges as does
the Jura to the Alps in Europe.

(3) A region of fragmentary ranges, separated by desert
depressions, including the Nushki desert and Khārān.

Little of the centre and west of Baluchistān has yet been
geologically explored. In the table on the next page, most of
the formations occur indifferently in any of the three regions.
The upper Eocene igneous intrusions are, however, restricted
to the second and third regional types, and the Makrān group
to the coast of that name.

Along the Great Boundary Fault we find the true Himālayan
structure typically exhibited, the Siwālik series of strata rising
out of the gravel plain and dipping towards the mountains
beyond, as if underlying their mass. South of the Zhob valley
occurs a succession of curved ranges in four distinct zones or
belts, the first or outer one Siwālik, the second Eocene, the
third Jurassic, and the fourth or innermost Triassic.

Of the various geological groups the upper and middle Constitu-
Siwāliks have, so far, proved unfossiliferous. The lowermost tion of
strata of the lower Siwāliks contain fresh-water shells and geological
remains of mammalia. The upper and lower Nāri appear to groups.
be conformable with one another, the latter being well represented in the Bolān Pass. The Spīntangi is a massive pale-
coloured limestone that caps the scarp of the Kīrthar range.
It is the most important of the nummulitic limestones in
Baluchistān. Shaly intercalations occur here and there in the
Spīntangi limestone, and become gradually more abundant

[1] Though not locally recognized, this name has come into general use
in scientific literature. It includes the Marri and Bugti country and the
Districts of Sibi and Loralai.

[2] For an explanation of the Great Boundary Fault and a discussion of
the Himālayan structure, see *Manual of the Geology of India.*

towards its base, thus passing into the next underlying group called the Ghāzij. The Ghāzij beds are the only formations among all the rocks in Baluchistān that have as yet proved of economic importance, owing to the coal seams which they contain. When the thickness of the Ghāzij increases considerably, this group becomes very similar in appearance to

Sedimentary rocks.	Approximate age.	Igneous rocks.
Recent and sub-recent alluvial and eolian formations.	Quaternary (Pleistocene and recent).	Recent and sub-recent volcanoes.
[Makrān group] Siwāliks { Upper.	Upper Miocene or Lowermost Pliocene.	
Middle. Lower.	Middle and Upper Miocene.	
Nāri { Upper. Lower.	Oligocene. Priabonian.	
Unconformity.	Bartonian.	Basic dykes. Intrusions of granite, diorite, and syenite.
Upper Kīrthar { Ghāzij and Spīntangi. Khojak shales (flysch facies).	Lutetian.	
Unconformity.	Thanetian to Lower Lutetian.	Intrusive, effusive, and stratified rocks of the Deccan trap formation.
Cardita Beaumonti beds (with flysch).	Montian or Thanetian.	
Dunghān group (with flysch).	Upper Senonian.	
Unconformity.	Neocomian to Lower Senonian.	
Belemnite beds.	Neocomian.	
Unconformity.	Callovian to Portlandian.	
Massive Limestone.	Bajocian to Callovian.	
Shales and Limestones.	Lias.	
Oolitic Limestone.	Rhaetic.	
Shales and Limestone.	Upper Trias.	
Unconformity.	Lower Permian to Upper Trias.	
Shales and Limestones.	Permo-carboniferous.	

the Khojak shales, the age of which is well established by the presence of nummulites contained in calcareous shales and massive dark bituminous limestone towards the base of the group. *Cardita Beaumonti* beds occur in the form of detached patches in many parts. The exposures of the Dunghān series are of limited extent. This consists of an extremely variable series of shales and limestone often merging into the flysch facies. Below comes a group of shales of Neocomian age containing innumerable specimens of belemnites, known as Belemnite beds. They rest on the massive limestone of which all the more conspicuous peaks in Baluchistān are composed. The triassic shales and limestones, forming an extensive outcrop south of the Zhob valley, are profusely injected by great intrusive masses of coarse-grained gabbro, often altered into serpentine, and innumerable dolerite or basalt veins and dykes of the Deccan trap age. To this period also belong many of the igneous rocks, both intrusive and eruptive, which occur abundantly in all three of the regional types mentioned above. A second group of igneous rocks is represented by deep-seated intrusions, without any connexion with volcanoes; it belongs to the second and third regional types, where it forms the granite and diorite of the Khwāja Amrān, the augite-syenite and augite-diorite of the Rās Koh, and the hornblende-diorite of Chāgai. Owing to the absence of rain, the materials formed by the disintegration of the mountains are not removed by rivers, but form immense deposits which are, therefore, of enormous depth. The desert plains represent vast areas of subsidence in regions once occupied by inland seas or lakes.

The flora of the plains and lower highlands resembles in Botany [1]. general aspect the vegetation of Western Rājputāna and the adjoining parts of the Punjab. Trees and herbs are conspicuously absent; and the bare stony soil supports a desolate jungle of stunted scrub, the individual plants of which are almost all armed to the leaf-tip with spines, hooks, and prickles of diverse appearance but alike in malignancy. A few, like the two first mentioned below, dispense with leaves altogether; and others, like *Boucerosia*, protect their fleshy branches with a hide-like epidermis. The commoner constituents of this illfavoured flora are *Capparis aphylla, Periploca aphylla, Boucerosia, Tecoma undulata, Acanthodium spicatum, Prosopis spicigera, Withania coagulans, Zizyphus Jujuba, Salvadora oleoides, Calotropis procera, Caragana polyacantha,* three kinds of *Acacia, Leptadenia Spartium, Taverniera Nummularia, Physorhynchus brahuicus,* and *Alhagi camelorum.* In low-lying parts where water is available *Tamarix articulata* and *T. gallica* are found. Here and there *Euphorbia neriifolia* and the dwarf-palm (*Nannorhops Ritchieana*) occur, the latter often in great quantities. The herbaceous vegetation is very scanty, consisting of such plants as *Aerua javanica, Pluchea lanceolata, Fagonia arabica, Mibulus alatus,* and *Cassia obovata;* near water *Eclipta erecta;*

[1] From notes by Major D. Prain, Director, Botanical Survey of India.

and as weeds of cultivation *Solanum dulcamara*, and *Spergularia*. Two species akin to *Haloxylon*, *Suaeda vermiculata* and *Salsola foetida*, abound on saline soil. *Panicum antidotale* is the most important grass, but *Eleusine flagellifera* and a species of *Eragrostis* are also abundant.

In the upper highlands the flora is of a quite different type. The long flat valleys have, for the greater part of the year, a monotonous covering of *Artemisia* and *Haloxylon Griffithii*, diversified, where there are streams, with tamarisks and species of *Salsola*, *Arenaria*, *Halocharis*, &c. On the surrounding hills, up to an elevation of 7,000 feet above sea-level, are to be found species of *Acantholimon*, *Acanthophyllum*, *Salvia*, *Amygdalus*, *Spiraea*, *Gentiana*, *Eremostachys*, and *Campanula*. Pistachio trees, associated with ash, wild olive, and daphne, are also common. Myrtle is occasionally found in the valleys. At higher elevations *Juniperus macropoda* and *Prunus eburnea* are abundant. Other plants common at these altitudes are *Lonicera*, *Caragana ambigua*, *Berberis*, *Cotoneaster nummularia*, *Spiraea brahuica*, *Rosa Beggeriana*, *Salvia cabulica*, *Berchemia lineata*, *Viola kunawarensis*, *Leptorhabdos Benthamiana*, and two varieties of *Pennisetum*. With the coming of spring a host of bulbous and other herbaceous plants, which have lain hidden throughout the winter, send forth leaves and flowers and for a few weeks make the valleys and hill-sides gay with blossoms of divers hues. They include four varieties of *Iris*, *Hyacinthus glaucus*, *Tulipa chrysantha*, *Tulipa montana*, *Fritillaria*, *Eremurus persicus*, *Cheiranthus Stocksianus*, *Campanula Griffithii*, *Delphinium persicum*, several species of *Alyssum*, and many species of *Astragalus*. In swampy grass lands spring up *Ononis hircina*, *Ranunculus aquatilis*, *Lotus corniculatus*, *Plantago major*, and *Eragrostis cynosuroides*. The weeds of cultivation include *Adonis aestivalis*, *Hypecoum procumbens*, *Fumaria parviflora*, *Malcolmia africana*, *Sisymbrium Sophia*, *Lepidium Draba*, *Malva rotundifolia*, *Veronica agrestis*, and many others. This many-coloured carpet of flowers endures for all too brief a season, for, under the intolerable heat of the summer sun, it speedily shrivels and disappears.

Zoology.

The fauna has never been completely studied. In the higher hills are to be found the mountain sheep (*Ovis vignei*) and the *mārkhor* (*Capra falconeri*). The latter, which is of the Kābul and Sulaimān varieties, lives a solitary life in the glens and cavities of the mountains, while the mountain sheep wanders on the lower slopes. In the lower highlands the *mārkhor* is replaced by the Sind ibex (*Capra aegagrus*). The leopard (*Felis pardus*) is frequently seen, and the black bear (*Ursus torquatus*) is found here and there. The Asiatic wild ass (*Equus hemionus*) haunts the deserts of Khārān and Nushki. The Indian wolf (*Canis pallipes*) sometimes occurs in considerable numbers and does much damage to flocks. Several kinds of foxes are found, their skins being in some demand.

The characteristic game birds of the country are *chikor* (*Caccabis chucar*) and *sīsī* (*Ammoperdix bonhami*), which abound in years of good rainfall and afford excellent sport. Large flocks of sand-grouse pass through the country in the winter, and the tanks are frequented by many varieties of wild-fowl. A few woodcock are also to be found. Most of the birds of Baluchistān are migratory. Of those permanently resident, the most characteristic are the raven, frequent everywhere; the lammergeyer, for which no place is too wild; and the golden eagle. Among the visitors the most common are different species of *Saxicola*, headed by the pied chat, and several kinds of shrikes which appear in spring in large numbers. Sea-birds are numerous along the coast.

Reptiles include the tortoise, several *genera* of lizards, of which the species *Phrynocephalus* is the most common, and the skink. Eleven *genera* of snakes have so far been discovered, the most numerous in species being *Zamenis*, *Lytorhynchus*, and *Distira*. They also include *Eristocophis macmahonii*.

The coast swarms with fish and molluscs, the former including sharks, perch, cat-fish, herrings, yellow-tails, and pomfrets.

The study of insects has been confined almost entirely to Quetta-Pishīn. Two species of locust are among the most conspicuous, and dragon-flies are common, as also are bees, wasps, &c. The latter include both Indian and European species, and many of them have been described by Russian naturalists. Butterflies are scarce, but moths are fairly numerous. Ants are found plentifully, but few species have been recognized. Sand-flies are common, and few persons escape their irritating attentions. Among the lesser-known classes of insects may be mentioned cicadas, which sometimes appear in vast numbers, and *Argas persicus*, so noxious to human beings. Plant-lice do great damage to many of the trees.

In a country which includes such varied natural divisions, Climate differences of climate are varied and extreme. It is temperate and temor otherwise in proportion to local elevation above sea-level. perature. Climatic conditions similar to those of Sind prevail in the plains and lower highlands, but in the upper highlands the seasons of the year are as well marked as in Europe. Owing to the proximity of the hills the heat of the plains in summer is probably even greater than that of Jacobābād, where the mean temperature in July is 96°. ' O God, when thou hadst created Sibi and Dādhar, what object was there in conceiving hell?' says the native proverb. In this part of the country also the deadly simoom is not infrequent. During the short cold weather, on the other hand, the climate is delightful. In the upper highlands the heat is never intense, the mean temperature at Quetta in July being only 79°. Except at Chaman, the diurnal range is highest in November. In winter the thermometer frequently sinks below freezing-point; snow falls and icy winds blow. The following table indicates the average temperature at places for which statistics are available. The figures for Jacobābād, which is in Sind but lies close to the border of Baluchistān, have been inserted as typical of the conditions in the Kachhi plain :—

	Height of observatory above sea-level.	January.		May.		July.		November.	
		Mean.	Diurnal range*.	Mean.	Diurnal range*.	Mean.	Diurnal range*.	Mean.	Diurnal range*.
	Feet.								
Chaman .	4,311	43.2	18.1	79.6	27.7	88.3	26.6	57.6	24.5
Quetta . .	5,502	40.0	21.8	67.8	31.4	78.7	27.9	48.7	32.7
Jacobābād .	186	58.1	29.8	94.7	33.1	96.2	23.9	69.1	35.8

* Average difference between maximum and minimum temperature of each day.

Winds.   The mountainous character of the country affects the direction and force of the winds. These partake largely of the character of draughts traversing the funnel-like valleys and finally striking the hills, where they empty the vapour that they carry. The north-west wind, known to the natives as *gorīch*, blows constantly. It is bitterly cold in winter and, in the west, scorching in summer. The west of Chāgai is subject to the effect of the wind so well known in Persia as the *bād-i-sad-o-bīst-roz*, or 120 days' wind.

Rainfall.   Baluchistān lies outside the monsoon area and its rainfall is exceedingly irregular and scanty. Shāhrig, which has the largest rainfall, can boast of no more than 11¾ inches in the year. In the highlands few places receive more than 10 inches, and in the plains the average amount is about 5 inches, decreasing in some cases to 3. The plains and the lower highlands receive their largest rainfall in the summer, and the upper highlands in the winter from the shallow storms advancing from the Persian plateau. In the former area the wettest month is July; in the latter February. The table below shows, for eight to twelve year periods ending in 1901, the average rainfall received at principal places :—

	January.	February.	March.	April.	May.	June.	July.	August.	September.	October.	November.	December.	Total.
Chaman . .	1·48	1·78	1·18	0·51	0·11	0·07	0·11	...	...	0·03	0·35	1·08	6·70
Quetta . .	1·98	2·12	1·78	1·06	0·47	0·17	0·93	0·56	0·11	0·08	0·34	0·92	10·52
Sibi . .	0·53	0·32	0·27	0·09	0·29	0·32	1·26	1·19	0·24	...	0·14	0·51	5·16
Loralai . .	0·98	1·13	1·06	0·47	0·80	0·41	1·72	0·99	0·11	0·06	0·15	0·49	8·37
Fort Sandeman	0·62	1·04	1·57	0·78	0·71	0·98	2·49	1·15	0·07	0·05	0·25	0·30	10·01

The conformation of the surface of the country renders much damage from floods impossible, but the vast volumes of water which occasionally sweep down the river channels some-times cause harm and loss of life. The floods are generally very sudden, and have been known to rise to a great height in the Nāri. The only known cyclone was that which visited Las Bela in June, 1902, destroying many cattle. Earthquakes are common everywhere and are frequently continuous. They sometimes cause much damage. A large earthquake crack has been traced for no less than 120 miles along the Khwāja Amrān and Sarlath ranges, and near this range of hills a disas-trous earthquake occurred on December 20, 1892. Storms, floods, cyclones, and earth-quakes.

One of the most striking facts in the history of Baluchistān is that, while many of the great conquerors of India have passed across her borders, they have left few permanent marks of their presence. Macedonian, Arab, Ghaznivid, Mongol, Mughal, Durrāni, all traversed the country, and occupied it to guard their lines of communication, but have bequeathed neither buildings nor other monument of their presence. History.

The earliest-known mention of part of Baluchistān is in the Avesta, the *Vara Pishin-anha* of which is undoubtedly identifiable with the valley of Pishīn. In the *Shāhnāma* we have an account of the conquest of Makrān by Kai Khusrū (Cyrus), and the Achaemenian empire which reached its farthest limits under Darius Hystaspes included the whole of the country. Among Greek authors Herodotus gives us little information about Baluchistān. He merely mentions Paktyake, which has been identified with the country of the Pakhtūns or Afghāns. It is to Strabo that we owe the best account, and from his writings we are able to identify the localities into which Baluchistān was distributed in ancient geography. On the north-east, and probably including all the upper highlands, The country as described in ancient geography.

was Arachosia; directly west of it was Drangiana; to the south lay Gedrosia; while in the Ichthyophagoi, Oraitai, and Arabies or Arabii are to be identified the fishermen of the Makrān coast, the inhabitants of Las Bela, and the people of the Hab river valley respectively.

Alexander the Great. The Graeco-Bactrians, &c. Alexander's retreat from India led him through Las Bela and Makrān, while a second division of his army under Crateros traversed the Mūla Pass, and a third coasted along the shore under Nearchus. Alexander's march and the sufferings of his troops are graphically described by Arrian. After Alexander's death, Baluchistān fell to Seleucus Nicator, and later on passed from his descendants to the Graeco-Bactrian kings who ruled also in Afghānistān and in the Punjab. Between 140 and 130 B. C., they were overthrown by a Central Asian horde, the Sakas, who passed along to the valley of the Helmand. About this time Buddhism, of which many traces are still to be found, flourished in Baluchistān. The empire of the Sassanians which followed expanded slowly towards the east, and Baluchistān was not conquered till the time of Nausherwān (529–577 A.D.).

The Rai dynasty and the Arab invasions. Henceforth the suzerainty over the petty rulers of Baluchistān alternated between east and west. In the fourteenth year of the Hijra (635–6) Rai Chach marched from Sind and conquered Makrān. The Rai dynasty at the same time appear to have extended their dominions to the north towards Kandahār. The Arabs reached Makrān as early as the year 643. The parts of Baluchistān which subsequently became best known to them were Turān (the Jhalawān country), with its capital at Khuzdār, and Nūdha or Būdha (Kachhi). Their power lasted till towards the end of the tenth century; for when Ibn Haukal visited India for the second time about 976, he found an Arab governor residing in Kaikānān (probably the modern Nāl) and governing Khuzdār.

The Ghaznivids. Shortly afterwards Baluchistān fell into the hands of Nasīr-ud-dīn Sabuktagīn; and his son, Mahmūd of Ghazni, was able to effect his conquests in Sind owing to his possession of Khuzdār. From the Ghaznivids it passed into the hands of the Ghorids and, a little later, was included in the dominion of Sultān Muhammad Khān of Khwārizm (Khiva) in 1219.

The Mongols. About 1223 a Mongol expedition under Chagatai, Chingiz Khān's son, penetrated as far as Makrān. A few years later Southern Baluchistān came under the rule of Sultān Altamsh of Delhi, but it appears soon to have reverted to the Mongols. The raids organized by Chingiz Khān have burned deep into the memory of Baluchistān. From Makrān to the Gomal, the Mongol (known to the people as the Mughal) and his atrocities are still a byword in every household.

Henceforth the history of Baluchistān centres round Kanda- *Migration* hār, and it was from this direction that in 1398 Pīr Muhammad, *of the* *Baloch* the grandson of Tīmūr, was engaged in reducing the Afghāns *and rise* of the Sulaimān mountains. Local tradition asserts that Tīmūr *of the* *Arghūn* himself passed through the Marri country during one of his *power.* Indian expeditions. The succeeding century is one of great historical interest. The Baloch extended their power to Kalāt, Kachhi, and the Punjab, and the wars took place between Mīr Chākar the Rind and Gwāhrām Lāshāri which are so celebrated in Baloch verse. In these wars a prominent part was played by Mīr Zunnūn Beg, Arghūn, who was governor of north-eastern Baluchistān under Sultān Husain Mirza of Herāt about 1470. At the same time the Brāhuis had been gradually gaining strength, and their little principality at this time extended through the Jhalawān country to Wad. The Arghūns shortly afterwards gave way before Bābar. From 1556 to 1595 the country was under the Safavid dynasty. Then it fell into the hands of the Mughals of Delhi until 1638, when it was again transferred to Persia.

We have an interesting account of Baluchistān in the *Ain-i-* *Baluchis-* *Akbarī.* In 1590 the upper highlands were included in the *tān under* *Akbar.* *sarkār* of Kandahār, while Kachhi was part of the Bhakkar *sarkār* of the Multān *Sūbah.* Makrān alone remained inde- pendent under the Maliks, Buledais, and Gichkīs, until Nasīr Khān I of Kalāt brought it within his power during the eighteenth century.

From the middle of the seventeenth century Baluchistān *Nādir* remained under the Safavids till the rise of the Ghilzai power *Shāh and* *Ahmad* in 1708. The latter in its turn gave way before Nādir Shāh, *Shāh.* who, during the first part of the eighteenth century, made several expeditions to or through Baluchistān. Ahmad Shāh, Durrāni, followed ; and thenceforth the north-eastern part of the country, including almost all the areas now under direct adminis- tration, remained under the more or less nominal suzerainty of the Sadozais and Bārakzais till 1879, when Pishīn, Duki, and Sibi passed into British hands by the Treaty of Gandamak.

Meanwhile the whole of Western Baluchistān had been con- *The rise* solidated into an organized state under the Ahmadzai Khāns *of the* *Ahmadzais* of Kalāt. All tradition asserts that the former rulers of Kalāt *of Kalāt.* were Hindus, Sewā by name. As Muhammadan dynasties held Baluchistān from about the seventh century, we must look to an earlier period for the date of the Sewās; and it is not improbable that they were connected with the Rai dynasty of Sind, whose genealogical table includes two rulers named Sihras. The Mīrwāris, from whom the Ahmadzais are descended, claim Arab origin. In their earlier legends we find them living at Surāb near Kalāt, and extending their power thence in wars with the Jats or Jadgāls. They then fell under the power of the Mughals; but one of their chiefs, Mīr Hasan, regained the capital from the Mughal governor, and he and his successors held Kalāt for twelve generations till the rise of Mīr Ahmad in 1666–7. It is from Mīr Ahmad that the eponym Ahmadzai is derived.

Authentic history now begins, and the following is a list of the rulers with approximate dates of their accession :—

1. Mīr Ahmad I, 1666–7.
2. Mīr Mehrāb, 1695–6.
3. Mīr Samandar, 1697–8.
4. Mīr Ahmad II, 1713–4.
5. Mīr Abdullah, 1715–6.
6. Mīr Muhabbat, 1730–1.
7. Mīr Muhammad Nasīr Khān I, 1750–1.
8. Mīr Mahmūd Khān I, 1793–4.
9. Mīr Mehrāb Khān, 1816–7.

10. Mīr Shāh Nawāz Khān, 1839.
11. Mīr Nasīr Khān II, 1840.
12. Mīr Khudādād Khān, 1857. From March, 1863, to May, 1864, the *masnad* was usurped by Khudādad Khān's cousin, Sherdil Khān.
13. Mīr Mahmūd Khan II, 1893. (The ruling Khān.)

Kalāt always a dependent State.

The rulers of Kalāt were never fully independent. There was always, as there is still, a paramount power to whom they were subject. In the earliest times they were merely petty chiefs; later they bowed to the orders of the Mughal emperors of Delhi and to the rulers of Kandahār, and supplied men-at-arms on demand. Most peremptory orders from the Afghān rulers to their vassals of Kalāt are still extant, and the predominance of the Sadozais and Bārakzais was acknowledged so late as 1838. It was not until the time of Nasīr Khān I that the titles of *Beglar Begi* (Chief of Chiefs) and *Wāli-i-Kalāt* (Governor of Kalāt) were conferred on the Kalāt ruler by the Afghān kings.

Gibbon's description of the history of Oriental dynasties, as 'one unceasing round of valour, greatness, discord, degeneracy, and decay,' applies well to the Ahmadzais. For the first 150 years, up to the death of Mīr Mahmūd Khān I, a gradual extension of power took place and the building up of a constitution which, looking at the condition of the country, is a marvel of political sagacity and practical statesmanship. A period of social ferment, anarchy, and rebellion succeeded, in which sanguinary revolts rapidly alternated with the restoration of a power ruthless in retaliation, until at length the British Government was forced to interfere.

As the Mughal power decayed, the Ahmadzai chiefs found themselves freed in some degree from external interference. The first problem that presented itself was to secure mutual cohesion and co-operation in the loose tribal organization of the state, and this was effected by adopting a policy of parcelling out a portion of all conquests among the poverty-stricken high-landers. Thus all gained a vested interest in the welfare of the community, while receiving provision for their maintenance. A period of expansion then commenced. Mīr Ahmad made successive descents on the plains of Sibi. Mīr Samandar extended his raids to Zhob, Bori, and Thal-Chotiāli, and levied an annual sum of Rs. 40,000 from the Kalhoras of Sind. Mīr Abdullah, the greatest conqueror of the dynasty, turned his attention westward to Makrān, while in the north-east he captured Pishīn and Shorāwak from the Ghilzai rulers of Kandahār. He was eventually slain in a fight with the Kalhoras at Jāndrihar near Sanni in Kachhi.

During the reign of Mīr Abdullah's successor, Mīr Muhabbat, Acquisition of Kachhi. Nādir Shāh rose to power, and the Ahmadzai ruler obtained through him in 1740 the cession of Kachhi, in compensation for the blood of Mīr Abdullah and the men who had fallen with him. The Brāhuis had now gained what highlanders must always covet, good cultivable lands ; and, by the wisdom of Muhabbat Khān and of his brother Nasīr Khān, certain tracts were distributed among the tribesmen on the condition of finding so many men-at-arms for the Khān's body of irregular troops. At the same time much of the revenue-paying land was retained by the Khān for himself.

The forty-four years of the rule of Nasīr Khān I, known to Nasīr Khān I. the Brāhuis as 'The Great,' and the hero of their history, were years of strenuous administration and organization inter-spersed with military expeditions. He accompanied Ahmad Shāh in his expeditions to Persia and India, while at home he was continuously engaged in the reduction of Makrān, and, after nine expeditions to that country, he obtained from the Gichkīs the right to the collection of half the revenues. A wise and able administrator, Nasīr Khān was distinguished for his prudence, activity, and enterprise. He was essentially a warrior and a conqueror, and his spare time was spent in hunting.

At the same time he was most attentive to religion, and
enjoined on his people strict attention to the precepts of the
Muhammadan law. His reign was free from those internecine
conflicts of which the subsequent story of Kalāt offers so sad
a record.

The reign of Nasīr Khān's successor, Mīr Mahmūd Khān,
was distinguished by little except revolts. In 1810 Pottinger
visited his capital and has left a full record of his experiences[1].

Mīr
Mehrāb
Khān.

The reign of Mīr Mehrāb Khān was one long struggle with
his chiefs, many of whom he murdered. He became dependent
on men of the stamp of Mullā Muhammad Hasan and Saiyid
Muhammad Sharīf, by whose treachery, at the beginning of the
first Afghān War, Sir William Macnaghten and Sir Alexander
Burnes were deceived into thinking that Mehrāb Khān was
a traitor to the British; that he had induced the tribes to
oppose the advance of the British army through the Bolān
Pass; and that finally, when Sir Alexander Burnes was
returning from a mission to Kalāt, he had caused a robbery
to be committed on the party, in the course of which an
agreement, which had been executed between the envoy and
the Khān, was carried off. This view determined the diversion
of Sir Thomas Willshire's brigade from Quetta to attack Kalāt
in 1839, an act which has been described by Malleson as
'more than a grave error, a crime[2].' The place was taken by
assault and Mehrāb Khān was slain. Shāh Nawāz Khān was
now appointed to succeed, with Lieutenant Loveday as
Political Officer. He was not, however, destined to occupy
the *masnad* for long. In the year 1840 a rebellion of the
Sarawān tribesmen caused his abdication, and Mīr Muhammad
Hasan, afterwards known as Mīr Nasīr Khān II, was placed
upon it.

Mīr Nasīr
Khān II.

By the efforts of Colonel Stacy, Mīr Nasīr Khān II was
induced to submit to the British Government, and was installed
by Major (afterwards Sir James) Outram at Kalāt in 1840.
Nasīr Khān at first acknowledged Shāh Shujā as the paramount power in Baluchistān; but subsequent events in Kābul
caused this undertaking to be annulled, and in 1854, as a
consequence of the European imbroglio with Russia, a formal
treaty, the first of those with Kalāt, was concluded with the
British Government. Quarrels, which had meanwhile broken
out between the Khān and the chiefs, led to Nasīr Khān

[1] Henry Pottinger, *Travels in Beloochistan and Sinde* (1816).
[2] Malleson, *History of Afghanistan* (1878).

raising a small body of mercenary troops, a measure which the chiefs naturally regarded as a serious encroachment on their powers.

Nasīr Khān II died, perhaps by poison, in 1857, and was succeeded by Khudādād Khān, then a mere boy. One of the first acts of the new ruler was to open fire with his guns on the chiefs who lay encamped near the city of Kalāt ; and, from this time till 1876, the history of Kalāt contains little but one long chronicle of anarchy, revolt, and outrage, in which there were seven important and many minor rebellions. In March, 1863, through the machinations of Mullā Muhammad Raisāni, Sherdil, the Khān's cousin, attempted his assassination, but only succeeded in wounding him. A general insurrection ensued ; Sherdil Khān was declared ruler and Khudādād Khān retired to the frontier. Mullā Muhammad now joined the other side, and the Khān regained the *masnad* in 1864. Revolt after revolt followed, until an attempt was made by the Commissioner in Sind to arbitrate between the parties in 1873. It proved abortive, and Major Harrison, the British Agent, was thereupon withdrawn and the Khān's subsidy was stopped.

Mīr Khudādād Khān.

At this juncture, Sir Robert (then Major) Sandeman appeared on the scene. His first mission to Kalāt in 1875 was not entirely successful ; and, immediately after its departure from the capital, Nūr-ud-dīn, the Mengal chief, with many of his followers, was slain by the Khān owing, it was alleged, to a plot against the latter's life. But a few months later Major Sandeman was again on the spot, accompanied by a large escort. By his tact and firmness, the Mastung agreement, the Magna Charta of the Brāhui confederacy, was drawn up on July 13, 1875, and read out formally in Darbār. An account of its provisions is given below in the section on Native States.

The Mastung settlement.

To make the influence which had been thus acquired effective for the future, the British Government now accepted the responsibility, as the paramount power, of preserving the peace of the country, and a fresh treaty was concluded with the Khān in December, 1876. In the following year Sir Robert Sandeman was appointed Agent to the Governor-General, and Quetta was permanently occupied.

Occupation of Quetta.

The rest of the story of Kalāt is soon told. During Sir Robert Sandeman's lifetime, no serious revolts occurred, and, in spite of the waywardness of Khudādād Khān, he was always treated by the Agent to the Governor-General with

Abdication of Mīr Khudādād Khān.

the greatest courtesy and consideration. In March, 1893, the *mustaufi* or chief accountant, with his father, his son, and a follower, were murdered by the Khān's orders. The Khān appears to have suspected the *mustaufi* of treachery, and alleged that the latter had made an attempt on his life. Khudādād Khān's abdication was subsequently accepted by the Government of India in favour of his son, Mīr Mahmūd, the present Khān. Mīr Khudādād Khān was shortly afterwards removed with his second and third sons to Loralai, and is now (1906) living in Pishīn.

Mīr Mahmūd Khān II.
The reign of the present Khān has been distinguished by few events of importance. In 1897 the wave of unrest, which passed down the frontier, made itself felt in Baluchistān, where a movement among the Sarawān chiefs, which might have had serious consequences, was averted by the arrest and imprisonment of two of the ringleaders. In the same year an outbreak occurred in Makrān, and British troops engaged the Makrān rebels at Gokprosh in January, 1898, the ringleader with many of his followers being slain. Another outbreak occurred in Makrān in 1901, which was also put down by British troops by the capture of Nodiz fort.

History of Las Bela.
Little need be said here of the history of Las Bela. Although nominally subject to Kalāt, whose ruler could call on it for an armed force when required, and claimed the right of control over the amount of the dues levied on goods in transit from Sonmiāni, its chiefs (Jāms) have always asserted and maintained a greater or less degree of independence, their position being strengthened by matrimonial alliances with the Khāns of Kalāt. Since the advent of the British the bond of connexion between the two States has been almost entirely severed.

The British as paramount power in Baluchistān.
The first Afghān War.
The political connexion of the British Government with Baluchistān commences from the outbreak of the first Afghān War in 1839, when it was traversed by the Army of the Indus and was afterwards occupied until 1842 to protect the British lines of communication. The districts of Kachhi, Quetta, and Mastung were handed over to Shāh Shujā-ul-mulk, and political officers were appointed to administer the country and organize a system of intelligence. Continual trouble occurred with the tribes in Kachhi; and in 1840 Kahān, in the Marri hills, was occupied by Captain Lewis Brown. Here he was besieged for five months by the Marris, who defeated a relieving force under Major Clibborn. In the meantime the garrison at Quetta was attacked by the

neighbouring tribes, and was also invested by the insurgents, who had raised Nasīr Khān II of Kalāt to the *masnad*. On the retreat of this gathering to Dādhar and its defeat by a British force, Lieutenant Loveday, the Assistant Political Agent at Kalāt, was murdered. The year 1841 opened auspiciously, but closed with the disaster at Kābul, an event which reacted on the Baluchistān tribesmen. Fortunately the country was in charge of a man of brilliant abilities, Sir James Outram; and all remained quiet while General England's column was pushed up the Bolān Pass to Quetta at the beginning of 1842, only to be defeated in the unfortunate affair of Haikalzai in Pishīn. Then began the withdrawal from Afghānistān; the districts which had been assigned to Shāh Shujā were handed over to the Khān of Kalāt, and Quetta was finally evacuated in October.

In 1845 General Sir Charles Napier led a force of 7,000 men against the Bugtis; but in spite of the assistance given by their enemies, the Marris, the operations were only a qualified success. The charge of the Upper Sind Frontier and of Baluchistān devolved on Captain (afterwards General) John Jacob from the beginning of 1847, and he held it till his death in 1858. Jacob's indefatigable energy and military frontier methods belong to the history of Sind rather than of Baluchistān, but his influence in Kalāt was very great, and it was he who negotiated the first treaty with that State in 1854. From 1856 to 1873 Political Agents were deputed to Kalāt, who were subordinate to the Political Superintendent in Upper Sind. *[General John Jacob's administration.]*

The founder of the Baluchistān Province as it now exists was Sir Robert Sandeman, who gave his name to a policy which has been aptly described as humane, sympathetic, and civilizing. Sandeman was the first to break down the close border system and to realize that the Baloch and Brāhui chiefs, with their interests and influence, were a powerful factor for good. His policy, in short, was one of conciliatory intervention, tempered with lucrative employment and light taxation. *[Sir Robert Sandeman.]*

Captain Sandeman, as he then was, first came into contact with the Marris and Bugtis as Deputy-Commissioner of Dera Ghāzi Khān in 1867; and, in consequence of the relations he then established and of his successful dealings with these subjects of the Khān of Kalāt, he was invited to take part in the Mithankot conference, which was held in February, 1871, between the representatives of the Governments of the *[Constitution of the Baluchistān Agency.]*

Punjab and Sind. The result of this conference was to place Sandeman, in his relations with the Marris and the Bugtis, under the Political Superintendent of the Upper Sind Frontier, and to cause the extension of the system of employing tribal horsemen, with the object of maintaining friendly communications with the tribes. Shortly afterwards matters in Kalāt went from bad to worse; the missions of 1875 and 1876 already referred to took place, and Baluchistān became a separate Agency directly under the Governor-General.

The
second
Afghān
War.

The importance of the position which had been acquired on the frontier was soon to be illustrated. At the end of 1878, the second Afghān War broke out, and troops were hurried forward to Kandahār along a line of communication which Sandeman's policy had rendered absolutely safe. At the close of the first phase of the war Sir Robert Sandeman accompanied General Biddulph's column, which had been deputed to open up the country between Pishīn and Dera Ghāzi Khān. At Baghao an engagement took place with the Zhob and Bori Kākars under Shāh Jahān, Jogezai, in which they were defeated. By the Treaty of Gandamak (May, 1879), Pishīn, Sibi, Harnai, and Thal-Chotiāli were handed over by Yakūb Khān to the British Government, on condition that the Amīr should receive the surplus revenues after payment of the expenses of administration. This treaty was afterwards abrogated by the massacre of the British Resident and his escort at Kābul and the deposition of Yakūb Khān. Then followed the second phase of the Afghān War and the British defeat at Maiwand in July, 1880. As a result of the renewal of military operations, some of the Afghān tribes within the Agency became restive and had to be subdued. An outbreak, too, occurred among the Marris which was put down by a small expedition. At the close of the war the retention of the areas ceded by the Treaty of Gandamak was decided on at Sir Robert Sandeman's strenuous instance.

Adminis-
tration,
1882 to
1892.

The ten years succeeding 1882 were years of administrative and organizing activity. Arrangements were commenced for the proper collection of the land revenue, irrigation schemes received attention, dispensaries were started, forests developed, and communications opened out in every direction. The strategical importance, too, of the western and north-eastern portions of the Province was fully realized. Two expeditions were made to Makrān, the first in 1883–4, during which the disputes between the Nausherwānis and the Khān of Kalāt were settled; and the second in 1890–1, when the question of the better administration of Makrān was taken up. On the north-east an expedition was made against the Zhob Kākars in 1884, which resulted in their submission. In 1886 Bori was taken over and the cantonment of Loralai founded. In 1887 the status of the Agent to the Governor-General was raised from that of a Resident of the second class to that of a Resident of the first class; the assigned districts were declared to be British territory, and the Agent to the Governor-General was appointed Chief Commissioner for them. The year 1889 saw an expedition to the Gomal through Zhob, when the District was occupied and the station of Fort Sandeman selected. At the end of 1890 another expedition took place under the command of General Sir George White, the object being the punishment of two of the Zhob chiefs who had been raiding the Zhob valley, and the chastisement of the Khidarzai section of the Largha Shirānis. The chiefs were not captured, but the Shirāni country was occupied without opposition. The position thus taken up enabled the Gomal to be effectively flanked, and the Shirānis and other tribes of the Takht-i-Sulaimān to be brought under control. Sir Robert Sandeman died at Bela in January, 1892, universally mourned.

He was succeeded by General Sir James Browne, who died Later in 1896. A period of consolidation and demarcation followed. develop-ments. Nushki was permanently leased in 1899, and in 1903 the lands irrigated by the Sind Canals, now known as the Nasīrābād *tahsīl*, were acquired from Kalāt on a perpetual lease. In the same year the Political Agent was withdrawn from Southern Baluchistān, and Las Bela was placed under the Political Agent of Kalāt; the Loralai District was formed by taking parts of the Zhob and the Thal-Chotiāli Districts, and the name of the latter was changed to Sibi. Thus, in a little more than a quarter of a century, security has replaced anarchy, and peace and plenty prevail now in tracts formerly given over to bloodshed and perpetual poverty.

Baluchistān offers a virgin field to the archaeologist, and one Archaeo-which is not altogether unpromising. Throughout the country logy. curious mounds occur, now deserted, but generally covered with masses of broken pottery, which will probably some day afford good ground for excavation. When the site of the present arsenal at Quetta was being prepared a statuette of Hercules was discovered. Mounds opened at Nāl and Māmā-tāwa in the Jhalawān country have yielded interesting finds of pottery. That found at the former place possesses striking

resemblances to pottery of the eighth century B. C. found in
Cyprus and Phoenicia and of Mycenaean technique. At Hinidān in the valley of the Hab river, at Sūrāb in the Jhalawān
country, and in Las Bela highly ornamented tombs of unknown
origin are to be seen, which afford evidence of a system of superterrene burial. The *gabrbands*, or embankments of the fire-
worshippers, which are common throughout the Jhalawān
country, are also of considerable interest, while some of the
underground water-channels, both round Quetta and near
Turbat in Makrān, indicate the possession of scientific skill
which is entirely unknown at the present day. North of Bela
lies the curious cave-city of Gondrāni, the cave-dwellings being
hewn out of the conglomerate rock. At Chhalgari in Kachhi
are indications of interesting Buddhist remains. Such finds of
coins as have been made from time to time render it clear that
all sorts of traders, from ancient times to the present, have left
traces of themselves along the routes leading from Persia to
India. Near Dabar Kot in the Loralai District several coins
of the time of the Caliph Marwān II, struck at Balkh in the year
of the Hijra 128 (A. D. 745), have been unearthed; and at
Khuzdār in the Jhalawān country Ghaznivid coins of interest
have been picked up, chiefly those of Ibrāhīm (1059 to 1099)
and Bahrām Shāh (1118 to 1152). The Koh-i-Taftān, which,
though not actually in Baluchistān, is close to the western
border, has yielded a find of considerable value in the shape of
Indo-Scythian coins, some of which are now deposited in the
British Museum. Punch-marked coins have been discovered
in Zhob, and coins of the Shāhis of Kābul in Khārān.

Population.
Although an attempt was made to obtain a rough enumeration of the population of some parts of Administered areas in
1891, it was not until 1901 that any systematic census was
carried out. This Census extended over 81,632 square miles, but
omitted Makrān, Khārān, and Western Sinjrāni. In the towns
and certain other selected places a synchronous enumeration
took place, but elsewhere estimates only were made. The
accuracy of the available figures is not therefore absolute. The
total population amounted to 810,746 persons. According to
a careful estimate made in 1903, the population of Makrān
amounts to about 78,000, and a similar estimate puts the
population of Khārān at 19,600. That of Western Sinjrāni
may be reckoned at about 6,000. The total population of
the Province is, therefore, about 915,000 persons. Areas
directly Administered have an area of 46,692 square miles
and an estimated population of 349,187, of whom 343,187

were actually enumerated. The population of the Native States and of the Marri and Bugti tribal areas (85,163 square miles) is estimated at 565,400, of whom 467,559 were counted at the Census. Detailed figures for the different localities will be found in Table I at the end of this article.

The density of population in the area covered by the Census amounts to less than ten persons per square mile. Including the areas for which estimates have now been obtained, the density falls to seven persons per square mile. The highest density is to be found in Quetta-Pishīn, with its large urban population and well-irrigated tracts, which possesses twenty-two persons to the square mile. In Chāgai, on the other hand, only one person per square mile is to be found. The number of persons per house in 1901 was 4·54.

About 95 per cent. of the total population enumerated dwelt Towns and in rural areas. No inducement exists in Baluchistān for the villages. indigenous inhabitants to collect into towns, and a general tendency is apparent among the people to avoid living together in large communities. This accounts for the paucity of towns, of which there are only six. All of them had garrisons in 1903, with the exception of Sibi and Pishīn, and they have sprung up since the British occupation. They contain a population almost entirely alien, which has accompanied the new rulers, either in service or for purposes of trade. Striking evidence of this is afforded by the fact that only 158 per thousand of the persons living in towns speak vernaculars of Baluchistān. Similarly, the villages are not only few in number (2,813, or one for every 47 square miles), but their size is small, and most of them contain less than 500 inhabitants. They are, as a rule, mere collections of mud huts, which are evacuated in summer when the cultivators encamp near their fields in blanket tents. The prevalence of the nomadic habit, to which reference will be made later, is one of the most remarkable features in the population. One of its results is that throughout the country small detachments, each of some half-dozen households, live together owning cattle, sheep, and goats, and moving from place to place for pasturage.

Owing to the doubtful accuracy of the figures obtained Growth of by the Census of 1891, no reliable comparative statistics popula-exist by which the increase in the population can be gauged. tion. The Census of 1901 showed an increase in some rural areas of 45 per cent., but part of this is probably due to better methods of enumeration. When we consider, however, that previous to

1876 the condition of affairs represented the 'ebb and flow of might, right, possession, and spoliation,' there can be little doubt that the increase of population since the British occupation has been considerable. In Quetta town, where the figures are reliable, an increase of 20 per cent. occurred in the decade.

Migration. The figures of migration in the *Report on the Census of India*, 1901, show a net loss to Baluchistān of 35,986 persons, the total of emigrants enumerated in India outside Baluchistān being 70,267 against 34,281 immigrants. Migration to and from Baluchistān is of two kinds : periodic and temporary. Nearly all the highland population of the country take part in the periodic migration—towards the plains in the autumn and towards the highlands in the spring. A distinction is observable between the migrations of the Afghāns and the Brāhuis. The Afghāns move far afield, and their object is generally commerce, the transport trade, or search for work as labourers. The Brāhuis, on the other hand, move in a more limited circle ; few of them care for commerce, while such transport as they do is confined to short distances. The work in which they particularly engage is harvesting and fuel-carrying. Many of them spend nearly the whole of the year in harvesting. In October and November they cut the rice in Sind ; this is followed by the *jowār*, and later by the spring wheat and barley. Then the heat drives them upwards until the highlands are reached in June. July and August are the Brāhui's months of rest, and in September he starts downward again.

Temporary emigration is chiefly confined to Afghāns and Makrānis. The former roam all over India, and even make their way so far afield as Chinese-Turkistān and Australia. Makrānis make good workmen, and leave their homes in search of labour. This temporary emigration is compensated by the large immigration. The immigrants constitute the security, the motive force, and the brains of the country. They are soldiers, clerks, merchants, and artisans ; but few of them settle permanently in the country, a fact sufficiently indicated by the very small proportion of women (18 per cent.) who are found among them. The majority of them are drawn from the Punjab, the United Provinces, and Sind.

Age statistics. No detailed record of age was attempted in 1901, but adults were distinguished from minors during the enumeration. It was found that in 100,000 males there were 66,053 adults and 33,947 children, while among 100,000 females there were 64,352 adults and 35,648 children. Children are thus proportionately less in number, and adults more numerous, than in India. In the towns and other places where the regular schedule was used, a synchronous statement of the ages of the normal population divided into age-periods indicates that the alien population in Baluchistān varies largely from the normal. Normal figures must naturally show a decreasing series of numbers at each age; but in Baluchistān, owing to the large alien population, the figures gradually rise till the maximum is reached in the case of males between twenty-five and thirty, and in the case of females between twenty and twenty-five.

Longevity among the indigenous tribesmen appears to be infrequent. Exposure, bad nutrition, hunger, and sickness affect the age of the population; and the principle of survival of the fittest must necessarily prevail among an uncivilized people such as is found in Baluchistān. A member of a tribe whose usefulness is affected by disease becomes a social outcast depending for his subsistence on charity; and when in hard times these sources are dried up, the impaired constitution quickly sinks. Of infirmities, blindness is common, probably owing to the dry and dusty climate, malnutrition, and excessive grain diet. Leprosy does not appear to be endemic, and insanity is rare. Infant mortality is undoubtedly high, owing to the unhealthy surroundings, want of proper nourishment, and exposure with which infant life has to contend. *Vital statistics.*

The disproportion of the sexes in the towns in 1901 was very great, there being only 260 women to every 1,000 men. The excess is greater in winter than in summer, as many women leave for their homes in India to avoid the former season. Among the indigenous population, numbering 762,039, a total deficiency in females of 50,901 was indicated, and this deficiency was uniform in Districts, tribes, and groups of different religious denominations. In every 1,000 Afghāns there were found to be 540 males and 460 females; among the Baloch the figures were 552 males to 448 females, and among Brāhuis 523 males to 477 females. The highest proportion of females is thus to be found among the Brāhuis and the lowest among the Baloch. That these figures are not far from the truth is indicated by the comparatively high bride-price paid by Afghāns, reaching Rs. 400 to Rs. 500, while among the Brāhuis it is much lower. *Sex statistics.*

Every tribesman marries as soon as he possibly can, but the payment of bride-price frequently makes bachelorhood compulsory till middle age. Polygamy is desired by all but attained by few. As a rule, marriage among the Muhammadan *Marriage customs and civil condition.*

population does not take place till puberty. Some of the
Kākar Afghāns have a curious custom permitting cohabitation
after betrothal. A Brāhui or Baloch will always endeavour to
marry a first cousin. The Afghāns, on the other hand, give their
daughters to the highest bidder without regard to relationship.
Among both Afghāns and Brāhuis a widow passes to the de-
ceased husband's brother. Divorce, though a simple process, is
infrequent. Adultery is punished among Baloch and Brāhuis
by the death of the parties, but Afghāns will generally salve
their honour for a consideration in money or kind. In the
area under a regular census, where, however, conditions are
wholly anomalous, there were in 1901, 4,632 unmarried, 4,839
married, and 529 widowers in 10,000 males ; while among
10,000 females 3,539 were unmarried, 5,626 were married, and
835 were widowed. As might be expected in a population
which is largely Musalmān, the proportion of widows is less
than in India proper. The marriage of children at an early
age is much less common, among both Hindus and Muhammadans, than in the neighbouring Province of the Punjab. Of
Hindus more than twice as many are married under fifteen
years than of Muhammadans.

Language.   The indigenous languages prevailing in Baluchistān are
Pashtū, Brāhui, Eastern and Western Baluchi, Jatki or Siraiki,
Jadgāli or Sindhi, Khetrāni, and Lāsi. The Dehwārs of Kalāt
and Mastung speak Dehwāri, a kind of bastard Persian. The
Loris or minstrels and blacksmiths have a curious jargon
called Mokaki. The language of correspondence is Persian.
Of Indian vernaculars spoken in 1901 in the areas where the
standard schedule was used, Punjābi was the most common
with 20,263 speakers. Urdū came next with 9,331, and
then Sindī with 3,305. There were 3,584 English-speaking
persons.

Generally speaking, Pashtū is spoken throughout the country
lying eastward of a line drawn from Sāngān near Sibi to Chaman. In the south-eastern corner of Loralai, Khetrāni, a
dialect of Lahnda or Western Punjābi, is prevalent. In
the Marri-Bugti country and in parts of Kachhi Eastern
Baluchi occurs. The cultivators of the last-named area speak
both Jatki and Jadgāli, the latter language being more widely
spoken and being also prevalent in Las Bela. Brāhui is spoken
throughout the Sarawān and Jhalawān countries, but only
extends up to about the 66th parallel of longitude, where it
meets Western Baluchi. Affinity of race is no criterion of
language. All Afghāns do not speak Pashtū, nor do all

Brāhuis speak Brāhui. Sometimes one section of a tribe talks Brāhui and another Baluchi.

Of the principal indigenous languages, Pashtū and Baluchi belong to the Indo-Aryan family, while Brāhui has been placed by the latest inquirer, Dr. Grierson, among the Dravidian languages. Baluchi has two main dialects, Eastern and Western. Western Baluchi, also called Makrāni, is more largely impregnated with Persian words and expressions than the Eastern dialect. A considerable body of literature has sprung up in this language. The soft southern dialect of Pashtū, as distinguished from the Pakhtū or northern dialect, is alone spoken in Baluchistān. Popular literature is entirely oral, commemorating events of local importance and relating stories of love and war. An account of BRĀHUI and its speakers will be found elsewhere (*post*, pp. 89-91).

The Meds, the Afghāns, and the Jats appear to have been the inhabitants of Baluchistān at the time of the Arab invasions. The Meds now, as then, live on the coast. The Afghāns still cluster round their homes at the back of the Takht-i-Sulaimān. The Jats, in spite of the influx of Brāhui and Baloch, to this day compose the cultivating classes of Las Bela and Kachhi; and some of the Kūrks, whose insolence led to the final subjugation of Sind by the Arabs, are still to be found in the Jau valley in the Jhalawān country. The indigenous races of chief importance at the present day are the Afghāns, Baloch, Brāhuis, and Lāsis. The Jat cultivators now form only a small minority; but many of them have undoubtedly been absorbed by the Baloch and Brāhuis. Among religious and occupational groups may be mentioned Saiyids, Dehwārs, and the indigenous Hindus who live under the protection of the tribesmen and carry on the trade of the country. The Afghāns, Baloch, and Brāhuis have been determined by Mr. Risley to be Turko-Irānians. Their stature is above the mean; complexion fair; eyes dark but occasionally grey; hair on face plentiful; their heads are broad and their noses of great length. Baloch hold the Marri and Bugti hills, and parts of Kachhi, where they mingle with the Jats. The Brāhuis occupy the great mountain band between Quetta and Las Bela, and in Las Bela we again have Jats called Lāsis. In Makrān many mixed races occur, which may be divided into two principal groups: the dominant races forming a small minority, and the races of aboriginal type known as Baloch, Darzādahs, &c. In the north-west of the Province the Baloch occur again, while in Nushki and the north-east of Khārān Brāhuis are numerous.

Races and tribes.

In the area where a regular census was taken in 1901, Brāhuis were found to be the strongest race, numbering 292,879. Afghans came next with 199,457, and, after a considerable drop, Baloch with 80,552. Jats numbered 66,746, and Lāsis 37,158. The numerical predominance of the Afghans and the insignificance of the Baloch are worthy of remark.

The Afghāns.

The Afghāns, or Pashtūns as they describe themselves, appear to have been living not far from their present abode in the time of Herodotus, if the identification of his Paktyake with Pakhtūns be accepted. Cunningham considers that they are also identifiable with the Opokien of Hiuen Tsiang in the seventh century. At the beginning of the eleventh century they had already spread southwards as far as Multān, for Mahmūd of Ghazni attacked them there. Subsequently two of their tribes, the Lodī and Sūri, gave rulers to the throne of Delhi. But while scattered groups were pushed out east and west to seek power and even empire, the nucleus of the race still remained in its ancient haunts; and to this day we find its elder representatives clustering round the Takht-i-Sulaimān, while others have made their way south to Sibi and as far north as Dīr, Swāt, and Bājaur. According to the Afghān genealogies, Kais Abdur Rashīd, the thirty-seventh in descent from Malik Tālūt (King Saul), had three sons, Gurgusht, Saraban, and Baitan. Among the descendants of Gurgusht we have Mando Khels, Bābis, Kākars, and Panis. The Saraban division is represented by the Tarīns, Shirānis, Miānis, and Barech; the descendants of Baitan can be identified in the Baitanis living across the Gomal. The most numerous and important indigenous Afghān tribes in Baluchistān are the Kākar (105,444), Tarīn (37,906), Pani (20,682), and Shirāni (7,309). The Kākars are to be found in largest numbers in the Zhob, Quetta-Pishīn, and Loralai Districts. The Tarīns have two main branches, the Spīn Tarīn and Tor Tarīn, of whom the former live in the Loralai and Sibi Districts and the latter in the Sibi and Quetta-Pishīn Districts. The Panis include both the Mūsā Khels of Zhob and the Panis of Sibi. The Shirānis live in close proximity to the Takht-i-Sulaimān. Of their two divisions, the Bargha, or upper Shirānis, alone occupy territory in Baluchistān. Numerous Ghilzais, nearly all of whom are nomads, occupy the country to the south of the Gomal river in winter. They are labourers, traders, and expert *kārez* diggers.

The Baloch.

Baloch tradition indicates Aleppo as their country of origin. The latest inquirer arrives at the conclusion that they are

Irānians[1]. The word *Baloch* means 'nomads' or 'wanderers,' and is coupled by Ibn Haukal and others with the word *Koch.* Whatever their original habitat, the Baloch had taken up a position in close proximity to Makrān early in the seventh century, and to this day many of their tribal names bear the impress of the localities which they occupied in Persian Baluchistān. Hence they made their way eastward until in the fifteenth century we find them settled in Kachhi. The tribes of importance are the Marris, the Bugtis, the Buledis, the Magassis, and the Rinds. Of these, the Rinds and Magassis have been enrolled in the ranks of the Brāhui confederacy; but the Marris and Bugtis appear, even in the palmiest days of the Ahmadzai rulers, to have been more or less independent.

Except in South-Western Baluchistān, where no tribal system appears to exist, Afghāns, Baloch, and Brāhuis are all organized into tribes, each having a multitude of subdivisions, clans, sections, and sub-sections. There is a distinction, however, between the constitution of an Afghān tribe and that of a Brāhui or Baloch tribe. Among the former the feeling of kinship is a bond of union far stronger than among the latter, with whom common blood-feud forms the connecting link. Theoretically, an Afghān tribe is constituted from a number of kindred groups of agnates; in a few cases only are small attached groups to be found which are not descended from the common ancestor. On the other hand, the Brāhui or Baloch tribe is a political entity, composed of units of separate origin, clustering round a head group known as the *Sardār Khel.* It is recruited on a definite system: a new-comer first shares in the good and ill of the tribe; later he obtains a vested interest in the tribal welfare by receiving a portion of the tribal lands; and his admission is sealed with blood by the gift of a woman in marriage. The tribe is organized and officered expressly for offence under an hereditary chief and headmen of groups. Among Afghāns the leader does not necessarily hold by heredity, for the individual has great scope for asserting himself; once, however, he has gained a position, it is not difficult for him to maintain it, provided he receives external support, and this is largely the secret of the Sandeman policy. *The tribal system.*

The Afghāns are tall, robust, active, and well formed. Their strongly marked features and heavy eyebrows give their faces a somewhat savage expression. The complexion is ruddy; the beard is usually worn short, as also is the hair. Their general bearing is resolute, almost proud. Courage is with them the *Physical characteristics.*

[1] M. L. Dames, *The Baloch Race* (1904).

first of virtues, but they are cruel, coarse, and pitiless. Vengeance is a passion. Their cupidity and avarice are extreme. The Baloch presents a strong contrast to his Afghān neighbour. His build is shorter, and he is more spare and wiry. He has a bold bearing, frank manners, and is fairly truthful. He looks on courage as the highest virtue, and on hospitality as a sacred duty. He is an expert rider. His face is long and oval, and the nose aquiline. The hair is worn long, usually in oily curls, and cleanliness is considered a mark of effeminacy. A Baloch usually carries a sword, knife, and shield. He rides to the combat but fights on foot. Unlike the Afghān, he is seldom a religious bigot.

Religion. Out of 810,746 persons enumerated in 1901, 765,368 were Muhammadans, 38,158 Hindus, 4,026 Christians, 2,972 Sikhs, and 222 'others.' In 100,000 of the population there were thus 94,403 Muhammadans, 4,706 Hindus, 497 Christians, 367 Sikhs, and 27 persons of other denominations. Of the total Christian population 3,477 were Europeans, 124 Eurasians, and 425 natives. Members of the Anglican communion were most numerous, numbering 2,857.

Islām. Islām and Hinduism are the only indigenous religions. The spread of Islām in Baluchistān probably occurred very early in the Muhammadan era. In practice, animistic beliefs and superstition rather than orthodox Muhammadanism prevail, and there is general ignorance of the tenets of the faith. Although they are now professed Sunnis, there are indications that the Baloch and Brāhuis have been much influenced by Shiah doctrines. Of sects, the Zikris or Dais of Makrān are the most interesting. They are the followers of a *Māhdi*, who is stated to have come from Jaunpur in India, and they believe that the dispensation of the Prophet Muhammad is at an end. While denying many of the doctrines of Islām, they have imitated others. They have constituted their Kaaba at Koh-i-Murād near Turbat, and thither all good Zikris repair on pilgrimage in the month of Zil Hij. They are very priest-ridden, and believe their *mullās* to be endowed with miraculous powers. At the same time their alleged incestuous practices appear to have been much exaggerated by their critics. They include nearly half the population of Makrān. The Taibs ('penitents') of the Kachhi are another small but curious sect.

Hinduism. Hinduism has been modified by its Musalmān surroundings. Hindus have little or no compunction in drinking from Musalmān water-skins, and some of them keep Musalmān dependants for domestic service. The Rāmzais of the Loralai District

afford a curious example of the assimilation by Hindus of
Baloch dress and customs.

Christian missions are endeavouring to gain a footing by Christian
giving medical aid and education. Branches of the Church missions.
Missionary Society and of the Church of England Zanāna
Mission have been opened at Quetta. The Province forms part
of the Anglican diocese of Lahore and of the Roman Catholic
arch-diocese of Bombay.

The majority of the indigenous population are dependent Occupa-
for their livelihood on agriculture, the provision and care of tions.
animals, and transport. An Afghān and a Baloch, as a rule,
cultivates his own land. The Brāhuis dislike agriculture, and
prefer a pastoral life. Their lands are, therefore, cultivated
through tenants who belong to professional agricultural groups.
A reminiscence of the slavery which existed in the country
before British occupation is to be found in a population of
servile origin numbering 22,304 in 1901. These servile dependants are happy and contented, and cases of ill-treatment
seldom occur. Women take a large part in all occupations.
Not only have they ordinary household duties to perform, but
they take the flocks to graze, groom their husband's horse, and
assist in cultivation. When a husband dies, his widow is
looked on as a valuable asset in the division of his property,
owing to the custom of demanding bride-price.

Meals are generally taken twice, at midday and in the even- Food.
ing. Flesh, milk, which is highly prized, and cheese in various
forms, with wheaten or *jowār* bread, are the chief constituents.
In the highlands a kind of 'biltong' is prepared in the winter
from well-fattened sheep, and is much relished. Onions,
garlic, and fresh asafoetida stalks are most used as vegetables.
On the coast rice and fish are eaten, while in Makrān dates
and dried fish form the staple diet of the people. Brāhuis and
Baloch never condescend to eat with their women folk.

The Afghān wears a loose tunic, baggy drawers, a sheet or Dress.
blanket, sandals, and a felt overcoat with the sleeves hanging
loose. His women wear a loose scarlet or dark-blue shift, with
or without wide drawers, and a wrapper over the head. The
Baloch wears a smock reaching to his heels and pleated at the
waist, loose drawers, and a long cotton scarf. His headdress is
wound in rolls round his head, generally over a small skull-cap.
The colour must be white or as near it as dirt will allow. His
wife's clothes resemble those of Afghān women, but must be
red or white.

Mat huts and black blanket tents stretched on poles are the Dwellings.

characteristic dwellings of the country. They are of various
dimensions, some being as much as 50 feet long by 10 feet
broad. They are generally about 4 feet high. The walls are
of matting, home-spun blankets, or stones laid in mud. The
dwelling is partitioned in the centre by a hurdle, on one side
of which live the family and on the other the flocks and herds.
At the back of the human dwelling are piled the felts and
quilts used for bedding. The remainder of the furniture consists of a wooden bowl or two, an earthen pot, a flat stone
griddle for baking, and a few skins for water and grain. Permanent dwellings are numerous only in those parts where they
are required for protection from the climate, or where there is
much cultivation. The house of a well-to-do person generally
consists of a courtyard with three rooms in a line. They
always face east or south, and consist of a storehouse, a winter
room, and a summer room. Outside, in the courtyard, are
a kitchen and a stable for cattle. Sometimes the houses are
double-storeyed, the lower part being used as a storeroom.
Cultivators of the poorer class merely have two rooms without
a courtyard. In the plains an open shelter, roofed with brush-
wood and supported on posts, is used in summer. In Las
Bela a peculiarity of the houses is the wooden framework,
generally of tamarisk ; there are no windows, but light and air
are admitted through a windsail in the roof.

Disposal of the dead. The usual Muhammadan mode of burial is in vogue. The
aperture of the grave is narrow at the top but broader near the
bottom. In some parts a corpse is never taken out through the
door, but the mat wall is broken down for purposes of removal.
If a person dies far from home the body is sometimes temporarily interred pending removal to its native place.

Amuse-ments. Field sports are the usual amusements of the people. They
indulge in racing, shooting, coursing, and tent-pegging. In-
door recreations among the Brāhuis and Baloch include
singing, dancing, and a kind of draughts. The Afghāns are
fond of marbles, prisoner's base, quoits played with a circular
stone, and a game like hunt-the-slipper. Ram and cock-
fighting are much admired, but their chief delight is in
dancing. Some ten or twenty men or women stand in
a circle, with a musician in the centre, and the dancers
execute a number of figures, shouting, clapping their hands,
and snapping their fingers. Wrestling after the European
fashion is common among the Afghāns and Jats. Brāhuis are
fond of trying their strength by lifting weights. Egg-fighting
is also of frequent occurrence.

Fairs are held throughout the country on the occasion of the Fairs.
Muhammadan festivals of Id-ul-Fitr and the Id-uz-Zuha, when
general rejoicings take place. Shrines are common and are
constantly visited. The best-known places of pilgrimage are
Hinglāj and Shāh Bilāwal in Las Bela, and Pīr Lākha
Lahrāni in Kachhi.

Among Brāhuis and Baloch, children are named on the Names and
sixth night after birth; among Afghāns, on the third day. titles.
Boys are named after the Prophet, according to the Muhammadan custom. The Brāhuis borrow names from trees, plants,
animals, &c. The word Khān is frequently added, out of
courtesy, to the names of men of good birth, and Bībī, Nāz,
Bāno, Bakht, Gul, Khātūn to those of women; a native
gentleman prefixes Mīr, thus: Mīr Yūsuf Khān. The first
child usually receives the name of the grandmother or grandfather as the case may be, a practice which causes much con-
fusion. In stating his name a man will generally add the
name of his tribe (*kaum*), and if questioned further, will always
give his section and sub-section also. The sectional name is
formed by adding the suffix *khel* or *zai* to an eponym for
Afghāns, *zai* and *āni* for Brāhuis, and *āni* for Baloch, thus:
Sanzar Khel, Ahmadzai, Aliāni. Permanent villages are usually
named after individuals, with the addition of *kili, kot* or *got,
kalāt,* or *shahr.* An encampment is called *halk* or *khalk.*

Agriculture, camel-driving, and flock-owning constitute the Agriculture.
occupations on which the majority of the population depend. General
The proportion of agriculturists to flock-owners is probably conditions.
about three to one. In many cases both agriculture and
pasture are combined. Previous to the advent of the British,
life and property were so insecure that the cultivator deemed
himself fortunate if he reaped his harvest; the fastnesses of
the hills, on the other hand, afforded shelter and safety to the
herdsman. The spread of peace and security has been accompanied by a marked extension of agriculture, which accounts
for the increase in numbers of the purely cultivating classes,
such as Lāngavs, Dehwārs, and Dehkāns. Some tribes are still
almost entirely pastoral, including the Marris and Bugtis; the
Sulaimān Khel, Nāsir, and other Afghāns; and many Brāhuis.

The cultivable area of the country in comparison with the Geological
total area is necessarily small; for, with the exception of the formation
plains of Kachhi, Las Bela, and the Dasht in Makrān, cultiva- and soil.
tion is confined to the limited area lying in the centre of the
valleys between the mountains. Here there is generally
a fringe of permanently irrigated land cut up into small

polygons, while towards the hills lie larger fields surrounded by embankments three or four feet high, by which the rainfall is caught as it descends from the gravel slopes bordering the valleys. In the centre are sometimes found bright red clays, many of them highly saliferous. Elsewhere, as in the great Thal plateau, the valleys consist of loess deposits, apparently formed by accumulations of wind-blown dust. In the plains the soil is generally loess mixed with alluvium. The admixture of moisture-bearing sand is the usual test applied by the cultivators to the quality of a soil.

System of cultivation. The scanty rain and snowfall, averaging between 6 and 7 inches, is nowhere sufficient to ensure cultivation without artificial assistance. The husbandman's return, therefore, is only assured where his cultivation is dependent on the *kārez* or underground water-channel, on springs, or on streams. All other cultivation is carried on by the artificial dams mentioned above. The areas under cultivation are thus divided into *ābi*, i.e. lands that are permanently irrigated ; and *khushkāba*, i.e. 'dry' crop lands which include also land subject to flood cultivation (*sailāba*).

Seed-time and harvest. The season for sowing the principal crops in the plains occurs in July and August, for at this time the rivers bring down the necessary moisture. In the upper highlands the dams are filled by the winter rainfall between December and March, when wheat and barley are sown. Here the system of cultivation in 'wet' crop differs from that in 'dry' crop areas. In the former the land is first watered and then ploughed, after which the seed is sown broadcast, and for further irrigation the fields are afterwards subdivided into small plots. In the latter, wheat, barley, and other spring crops are sown with the drill (*nāli*), and the seed depends for further moisture on the subsequent rainfall. In the plains, where the only important crop is *jowār*, the seed is everywhere sown broadcast after the ground has absorbed the flood moisture. There are two harvests, gathered in the spring and the autumn. The spring harvest, known to the Afghāns as *sparlai* or *dobe* and elsewhere as *sarav* or *arhāri*, is most important in the uplands, while the autumn harvest, known as *mane* in Pashtū and *sānwanri* or *āmen* elsewhere, takes its place in the plains.

Principal crops. Of spring crops, the most important is wheat. *Jowār* (*Andropogon sorghum*) is the chief autumn crop in the plains, and maize in the highlands. Dates constitute the sole crop of importance in Makrān. *Mūng* (*Phaseolus mungo*) and oilseeds in the plains and tobacco and melons in the hills also occupy

considerable areas. Of minor products may be mentioned
barley, gram, and beans among spring crops ; and rice, several
varieties of millet, gingelly, cotton, and a little indigo among
autumn crops. Lucerne yields from May to October.

Wheat is of several varieties, both red and white. The red Wheat.
is preferred for home consumption, but the white fetches better
prices in the market. In the highlands the seed can be sown
from the middle of September to the end of January. In the
plains wheat is not grown unless there are late floods towards
the end of August, affording moisture which is carefully
preserved for sowing as soon as the heat of summer subsides
in October. Both autumn and winter wheat are cultivated,
the former ripening in about nine months and the latter in
about five or six. In the highlands the crop is often grazed
down with sheep or goats up to February. The harvest
commences about June. The straw is carefully preserved
for fodder (*bhūsa*). The local varieties are hardy, but are
affected by sudden changes of climate and by much rain in
early spring.

Jowār is sown broadcast as soon as the lands which have *Jowār*.
absorbed the summer floods have been ploughed. If spring
floods occur, much *jowār* is cultivated for fodder, and the
same plants, if they receive water from the summer floods,
bear a good crop of grain. Many varieties occur, of which
those known as *thuri* and *thor* are most extensively cultivated.
Not only does *jowār* form the staple food-grain of the parts
where it is cultivated, but the fresh stalks contain saccharine
matter which is much relished. When dry, they constitute
excellent fodder (*karb*).

A large increase in the cultivation of melons, known locally Melons.
as *pālezāt*, has taken place. Both water-melons and sweet
melons are grown. Sweet melons are of two varieties : *garma*,
or summer melons ; and *sarda*, or autumn melons. *Garma*
melons are of several kinds—spotted, streaked, or white—and
are eaten fresh. It is a peculiarity of *sarda* melons that they
can be kept for several months. Those grown from imported
Kābul seed are considered the best. The cultivation of melons
has been much improved by the introduction of the *joia* system
from Kandahār. After the land has been ploughed, long raised
beds are formed, enclosed by an irrigation trench on either side
about one foot deep. The seed is sown along the edges of the
trench and, when small, the plant is carefully pruned. At
a later stage poor flowers are picked off and the young melons
are buried in the soil to avoid disease.

Dates.   Throughout Makrān the staple food is dates. Great attention is paid to the cultivation and care of the date-tree, and the
dates of Panjgūr are declared by Arabs to excel those of Basra.
Though all the trees belong to the species *Phoenix dactylifera*,
they are distinguished locally into more than a hundred kinds,
according to the weight, size, and quality of the fruit. All trees
are known either as pedigree trees (*nasabi*) or non-pedigree
trees (*kuroch*). Among the former the best varieties are *mozāti,
āp-e-dandān, haleni, begam jangi*, and *sabzo*. Fresh trees are
raised from offsets, and they produce fruit after three to eight
years, and continue to do so for three generations. The young
offsets must be carefully watered for the first year, and afterwards at intervals until their roots strike the moisture of the
subsoil. The date season is divided into three principal periods :
machosp, rang or *kulont*, and *āmen*. In *machosp* (March)
the artificial impregnation of the female date-spathes by
the insertion of pollen-bearing twigs from the spathe of the
male tree takes place. In the season of *rang* or *kulont* (June)
the colour first appears on the fruit, and there is general
rejoicing. The harvest (*āmen*) lasts from July to September,
when men and cattle live on little else but dates. The fruit is
preserved in various ways, the most common being by pressing
and packing in palm-leaf baskets (*laghati*). Better kinds are
mixed with expressed date-juice and preserved in earthen jars
known as *humb*. Owing to the excessive quantity of dates in
the diet of the people of Makrān, night-blindness is common.

Manure,   The use of manure is fully appreciated by the highland
out-turn, cultivators. Each one collects the sweepings from his cattle-
and   yard and carries them to his field, and in the neighbourhood
rotation. of the towns all the available manure is bought up. The
following table gives the average out-turn per acre of wheat
and *jowār* from experiments made in Administered areas :—

	Wheat. Cwt.	Jowār. Cwt.
Irrigated land, manured　·　·　·　·	13·3	13·6
Irrigated land, not manured　.　·　·　·	9·0	8·8
Land under flood-irrigation　.　·　·　·	13·6	12·9
'Dry' crop land　.　.　·　·　·　·	6·2	6·4

The fertility of land dependent on fl ood-irrigation is well
exemplified by these statistics. Manured and irrigated land
has been known to produce as much as 18 cwt. of wheat
per acre and 21 cwt. of *jowār*. No rotation is followed in
unirrigated lands, the silt brought down by floods being sufficient to ensure an excellent crop whenever there is enough

moisture for cultivation. Irrigated fields near homesteads, which can be manured and which therefore are generally cropped twice a year, are allowed to lie fallow in alternate years. In other irrigated lands three or four years of fallow are allowed after each crop.

Fruit is extensively grown in the highlands, and the export is Fruit. increasing. The principal kinds are grapes, the best of which are known as *lāl*, *sāhibi*, *haita*, and *kishmish*; apricots, mulberries, almonds, apples, pomegranates, peaches, nectarines, quinces, plums, and cherries. Much improvement has been effected by the introduction of fresh varieties. All kinds of English vegetables are grown. Excellent potatoes are produced and the cultivation is extending. The appointment of a Superintendent of Arboriculture and Fruit-growing was sanctioned in 1902, and large numbers of good fruit-trees are raised and distributed. An impetus has recently been given to mulberry cultivation by the inception of sericultural operations.

A test of the increase of cultivation which has taken place in Increase recent years is afforded by the returns of the Government share of cultivation and of revenue in kind. In 1879–80 the revenue in Sibi amounted new crops. to 6,575 maunds of wheat, while in 1902–3 it was 11,978 maunds; and between 1882 and 1895 the revenue in wheat from the Quetta *tahsīl* rose from a little more than 4,000 maunds to about 18,000 maunds. The cultivation of tobacco, potatoes, and oats has been recently introduced, and sugar-cane is making some way. Madder-growing, which was extensive at one time, has now ceased.

The implements of husbandry are primitive. The plough Imple- is similar to that used in India, but somewhat lighter owing to ments. the softness of the soil. A heavy log is used as a clod-crusher. For making large embankments a plank about 6 feet long and 2 feet wide, called *kenr*, is employed. Small embankments in irrigated lands are made with a large wooden spade (*dal*), which is worked by two men with a rope. Shovels of an improved pattern are now in use near Quetta. Mattocks have a flat blade. A four-pronged fork called *chār shākha* is used for winnowing and for cleaning straw. Efforts have been made, but without success, to introduce a plough worked by horses.

Though the Land Improvement and Agriculturists' Loans *Takāvi* Acts have not been formally applied, advances to cultivators advances. are made under executive authority. A special feature is the permission given to District officers to carry out improvements themselves with such advances. Owing to the backwardness of

the country, encouragement is given to applicants by the grant
of loans on easy terms or without interest. An annual sum of
Rs. 60,000 is allotted for the purpose, the advances being
ordinarily used for the construction of underground water-
channels, embankments, and wells. The people fully
appreciate the advantage of the system, and the political
effect has been excellent. The rate of interest usually charged
is 6¼ per cent. per annum. No difficulty has been found in
obtaining repayment. During the five years ending March,
1904, nearly 3 lakhs of rupees was advanced. About 2¼ lakhs
was outstanding at the end of that time.

Indebted-
ness of
the culti-
vators.

Except in the plains, the agriculturists largely finance one
another. The usual method of obtaining a loan is to mort-
gage the land with possession until both principal and interest
have been paid. Tenants-at-will are deeply involved in debt,
and live a hand-to-mouth existence. In the plains the Hindu
baniā plays an important part among the agricultural community. He keeps a running account with the cultivator on
the security of the latter's crop, and at each harvest receives
a part of the grain-heap as interest, with such amount as the
cultivator can spare towards the reduction of the principal.
The usual rate of interest is two annas in the rupee per annum,
but this rate is only allowed by the *baniā* to those cultivators
who give *mahtai*. This is a measure given by the cultivator
from his grain-heap at each harvest to induce the *baniā* to
advance sums at low interest. Cultivators who do not give
mahtai have to pay four annas per rupee.

Cattle.

A thickly built bullock, of small size and generally black
or brown in colour, is found in the hills and is well suited
to them. A pair fetch about Rs. 60 to Rs. 80. The bullocks
bred at Bālā Nāri in Kachhi, being suitable for agricultural,
siege-train, and army transport purposes, are much sought after
by dealers from the Punjab. They are of two distinct types.
The taller ones are 56 inches at the shoulder, white or fawn in
colour, with horns growing upwards and inwards. The other
type is smoky white, with black legs and neck, 42 to 48 inches
high at the shoulder, and with horns growing slightly upwards
and backwards. Both these kinds fetch good prices, a pair
selling for Rs. 100 or more.

Horses.

Baluchistān has long been noted for its breed of horses. As
early as the seventh century, Rai Chach of Sind took tribute in
horses from Gandāva. This reputation has ever since been
maintained; and in pre-British days the Huramzai Saiyids of
Pishīn and many of the Brāhui tribesmen were in the habit of

taking horses for sale so far afield as Mysore. Pedigree Baloch mares are still much prized, especially those of the *Hirzai* breed of Shorān. The best animals in the country are those bred round Mastung and at Jhal in Kachhi. They are big powerful animals with plenty of bone. Another good breed is found in Bārkhān, these horses being about 15 hands in height, with small well-bred heads and long slender well-arched necks. Their legs are small below the knee, but they are very hardy. The Government, therefore, found much excellent material for commencing horse-breeding operations, when it decided to introduce government stallions in 1884 under the superintendence of the Civil Veterinary department. The stallions are all Arabs or English and Australian thoroughbreds, the services of which are allowed free of charge to owners for mares which have been branded after inspection. As indicating the results hitherto attained, it may be noted that the Horse-breeding Commission in 1901 pronounced one of the classes of brood mares at the Sibi fair as good as anything to be seen in England. In 1903 the Army Remount Department took over charge of the operations. The country contained 1,276 branded mares in 1904, and 379 foals from government sires. In the same year 36 stallions were employed. Horses vary much in price. A tribesman can generally obtain a good mount for Rs. 100 to Rs. 150, but well-bred animals fetch Rs. 400 and more.

The heavy transport of the country is done entirely on camels. They are of the small hill-bred type, excellent over rocky ground but unable to stand the great heat of the plains of India. Their usual load is about 400 lb. They are bred chiefly in the Quetta-Pishīn and Zhob Districts, the Marri-Bugti country, Kachhi, Khārān, and Las Bela. As a rule transport animals are readily available, but the number varies in different parts of the country with the season of the year. The total number of camels in the country has been estimated at about 350,000, but this figure is probably above the mark. In some parts they are used for ploughing. Special arrangements have been made by Government to organize camels for transport purposes. The price of a transport camel varies from Rs. 60 to Rs. 80, and of a breeding camel from Rs. 50 to Rs. 60. *Camels.*

Donkeys are used by every nomadic household and are most useful animals. They frequently carry over 300 lb. and require little or no fodder besides what they can pick up on the march. The Buzdār breed, obtainable in the Loralai District, is the *Donkeys.*

best, and there are some good ones near Kalāt. To enable donkeys to breathe freely when going uphill it is usual for their nostrils to be slit soon after birth. An ordinary donkey fetches from Rs. 20 to Rs. 30; large donkeys from Rs. 60 to Rs. 80. Encouragement is now being given by Government to donkey-breeding on the same lines as to horse-breeding, and six donkey stallions were stationed in the country in 1904.

Sheep and goats. The sheep are of the fat-tailed variety, white, brown, and grey in colour. The white with black faces preponderate. A breed imported from Siāhband, near Kandahār in Afghanistān, is preferred for both meat and wool. Sheep are shorn twice, in spring and in autumn, producing 2 to 3 lb. of wool. The quality of the wool is coarse, and it comes to the market in very dirty condition. The goats are small and generally black. They are not very hardy. They yield about $1\frac{1}{2}$ lb. of hair, which is generally used at home for making blanketing, ropes, grain-bags, &c. Both sheep and goats are very numerous, and constitute much of the agricultural wealth of the country. The average price of a sheep is from Rs. 4 to Rs. 5. Goats fetch from Rs. 3 to Rs. 4.

Fodder. The question of fodder is one of the most difficult in Baluchistān, since no large quantity of grass exists in the greater part of the country, and horses and bullocks subsist chiefly on the straw of cereals. The best fodder available for horses is straw mixed with lucerne, but it is expensive. In the plains the stalks (*karb*) of *jowār* constitute almost the only fodder. The large herds of sheep and goats, which rove over the hills for six or seven months of the year, keep in excellent condition owing to the numberless small cruciferous and leguminous plants, which afford good pasturage. The goats also obtain grazing from the bush growth. Camels find abundant fodder in the salsolaceous plants, *Alhagi camelorum*, tamarisk, &c., and are fond of browsing on most of the trees. The best grazing tracts are to be found in the Loralai District.

Fairs. Horse and cattle fairs are held at Sibi in February and at Quetta in September. The former is chiefly a breeder's and the latter a dealer's fair. About 1,800 horses are brought to these shows, and prizes to the value of Rs. 6,000 are given. At the Quetta fair many Persian horses are brought for sale.

Cattle diseases. Cattle suffer considerably from diseases of the pulmonary organs owing to the cold. Foot and mouth disease also occurs occasionally. Mange in goats and camels is common. Camels also suffer from colic, rheumatism, fever, and cough. A gadfly causes some mortality in summer, and the cold induces pneumonia in winter. The Superintendent, Civil Veterinary department, Sind, Baluchistān, and Rājputāna, controls the operations of the department in the Province.

The majority of the crops depend either on permanent or Irrigation. flood irrigation, and those raised from rain-water are insignificant. Except *jowār*, *mūng*, and oilseeds, for which a single flooding of the land is sufficient, all other crops require further waterings to bring them to maturity. The sources of permanent irrigation are Government canals, underground water-channels (*kārez*), springs, and streams. Temporary irrigation is obtained by constructing embankments along the slopes of the hills, or by throwing large dams across the river-beds to raise the flood-water to the level of the surrounding country. In the highlands the two principal irrigation works are the Shebo and Khushdil Khān schemes in Pishīn. These 'minor' works have been constructed at a cost of nearly 17 lakhs, and irrigate annually about 6,000 acres. The return on the capital outlay is less than 1 per cent. in each case, but they have produced an excellent political effect in settling down the inhabitants. The revenue is levied in kind at one-third of the gross produce, which includes water-rate. In the plains, the Begāri and Desert Canals, which form part of the Upper Sind system, afford irrigation in the Nasīrābād *tahsīl* of the Sibi District. The assessment on the former is Rs. 2 per acre, of which R. 1 is reckoned as water-rate and R. 1 as revenue, and on the latter Rs. 2.8, of which the water-rate amounts to Rs. 1–8. A small cess of 6 pies per acre is also levied. The total area irrigated in 1902–3 was 105,962 acres, and the water-rate realized Rs. 1,27,404. Improvements are now under consideration for extending the area commanded by these canals, and a revision of the assessment is contemplated. Twenty-four artesian wells of moderate depth have been bored near Quetta. The Irrigation Commission (1903) considered that experimental borings in Baluchistān appeared to hold out more hope of securing an artesian supply of water at moderate depth than in any other part of India, and steps are being taken to experiment further on a larger scale.

The indigenous sources of irrigation in administered areas Indigenous include 1,803 springs, 496 *kārez*, 132 streams, and 76 wells. sources of supply. The *kārez* is an underground tunnel, driven into the great The inosculating fans which spread with a slope of 300 to 600 feet *kārez*. per mile from the mouths of the hill ravines into the valleys. These tunnels have a slope less than that of the surface and, acting as a subsoil drain, carry the water out to the surface.

The cost varies, according to their size and the soil in which they are excavated, from Rs. 2,000 to Rs. 8,000. Most of the wells are in the Nasīrābād *tahsīl*; they irrigate about five acres each, and their cost varies from Rs. 500 to Rs. 1,000. Perhaps the most interesting system of indigenous irrigation is that prevailing in Kachhi, where the cultivators, under an organized method of co-operation, construct annually immense earthen dams in the Nāri river for raising the water to the surface. A specially expert cultivator, known as the *rāzā*, is selected to superintend the work, and the cultivators living for many miles along the banks of the river are called in with their bullocks to construct the dam. The implement used is the wooden plank-harrow (*kenr*). Some of these dams are as much as 750 feet long, 180 feet broad at the foot, and 50 or 60 feet in height. Every village has to supply its quota of men and bullocks, or, should it fail to do so, has to pay a proportionate amount in cash. There are many of these dams in the Nāri; and in July and August, when the floods come, the upper dams are broken as soon as sufficient water for the area irrigable by each has been received. Still, much water runs to waste, and the scientific development of this indigenous system would probably result in a very large increase of cultivation.

Fisheries. The Makrān coast is famous for the quantity and quality of its fish, and the industry is constantly developing. It affords a considerable income to the States lying along the coast, as they generally take one-tenth of the fresh fish as duty. The fishermen are principally Meds and Koras. The fish are caught both with nets and with the hook and line. Large shoals of cat-fish (*gallo*) and of *kirr* (*Sciaena diacanthus*) appear off the coast towards the end of the cold weather, when they are surrounded and caught. On arrival on shore the air-bladders are extracted, and are eventually exported to England for the manufacture of isinglass. The fish are salted and used as food by the people of the country. They also form a large article of export to Bombay and Zanzibar. Sharks are prized especially for their fins, which fetch as much as R. 1 per lb. Fresh sardines are so plentiful that they not infrequently sell at the rate of forty for a pice (one farthing).

Classes of tenants. The bulk of the land is held by a cultivating class of peasant proprietors. The few tenants are almost all tenants-at-will. In local parlance such a cultivator is 'the husband of a slave girl,' for he can be replaced at his master's will. He can acquire no permanent rights, and is liable to ejectment after the crop has been harvested. Sometimes, when much labour has to be expended on the construction of embankments, a tenant retains possession so long as the embankment remains unbreached or for a given term of years. In Makrān a curious custom prevails, giving to a tenant-at-will a permanent alienable right in all date-trees which he may plant. Occupancy rights are seldom, if ever, acquired in irrigated and only occasionally in unirrigated lands. Where they have been so acquired, they usually represent compensation for the labour expended by the tenant on raising embankments.

As might be expected in a backward country in which crops Rates of are liable to great variations, rent consists in a share of the rent. grain-heap. Various systems are in vogue; but, as a general rule, the distribution in unirrigated lands is made on the principle of an assignment of a portion of the produce for each of the chief requisites of cultivation: the land, seed, bullocks, and labour. In irrigated lands a further, and proportionately large, share is assigned for the water. Certain services have also to be performed by the tenant, such as the supply of fuel and the transport of the proprietor's share of the produce. The position of the tenants on the whole is strong, since, owing to the inveterate laziness of the land-holding classes, there is a large demand for them and they can enforce their own terms.

No coolie class exists among the cultivating population; Wages. tenants-at-will perform the services mentioned above, while the household and agricultural work of men of means is done by their servile dependants. At harvest-time the workers, many of whom are women and children, receive a share of the grain-heap, generally one-twentieth. To shepherds are given their food, two changes of clothes, and a proportion of the lambs born during the year. The wages of village servants consist in a fixed measure from the grain-heap, or in a special share of water for irrigation. Coolie work proper is a peculiarity of the industrial centres which have grown up since the British occupation, and here a plentiful supply of labour from Makrān and Afghānistān is always to be found. As a navvy the Hazāra or Ghilzai Afghān is unrivalled. All domestic servants and skilled labourers come from India, chiefly from Sind and the Punjab.

Owing to the severity of the climate and the comparatively Rates of large amount of clothing and fuel required by the wage-earning wages. classes, wages throughout the highlands are higher than those usually prevalent in India. An unskilled labourer receives Rs. 11 to Rs. 15 a month; a skilled labourer, Rs. 20 to Rs. 45; mechanics, Rs. 35 to Rs. 90. The wages of domestic servants vary from Rs. 10 to Rs. 25 in European households; and from Rs. 6 to Rs. 12 with food among natives. The clerical wage rises from Rs. 20 for vernacular clerks to about Rs. 200 for those who know English. In a few special cases it is higher. A levy footman is generally paid Rs. 10 and a horseman Rs. 20 a month, for which sum the latter must maintain a mount. The opening of communications has not materially affected the wages of unskilled labour, but there has been a decrease in the earnings of artisans and clerks.

Prices.

Wheat is the staple food-grain in the highlands and *jowār* in the lower tracts. Firewood and chopped straw for fodder also form important items in domestic economy. Prices rule high when compared with those prevailing in India. The following table exhibits the average prices (retail) in seers[1] per rupee of staples at principal centres for the two quinquennial periods ending 1895 and 1900:—

Station.	Wheat.		*Jowār*.		Chopped straw fodder.		Firewood.		Punjab salt.		Country salt.	
	1895.	1900	1895.	1900.	1895.	1900.	1895.	1900.	1895.	1900.	1895.	1900.
Fort Sandeman	14	10½	15½	11½	59½	56	115	98¾	6	6½	14¾	11½
Loralai . .	17	11½	18¾	13¼	57½	58¾	75	79	6½	6¾	9½	9
Sibi . .	15¼	12½	22½	17½	109½	81¼	105¼	119	9¾	9½	14¾	13¼
Quetta . .	13¼	11	20¼	14½	48¾	45½	66	73	8¾	9¼	12	11

In 1903 the price of wheat at Fort Sandeman and Quetta was 11¾ seers per rupee, equivalent to about 17½ lb. for 1*s*., and at Loralai 14½ seers, equivalent to 22 lb. for 1*s*. The price of *jowār* in the plains at Sibi was 20½ seers per rupee, equivalent to 31 lb. for 1*s*. The purchasing power of the rupee in the case of the more important staples shows a marked decrease during the decade ending in 1900, but the period was one of abnormally bad agricultural conditions. Prices are affected largely by the seasons. They are always lower in the plains than in the hills.

Material condition of the people.

There has been a steady improvement in the standard of life throughout Baluchistān since the British occupation. This is more marked in the tracts under British administration than in the Kalāt State. Tea is now becoming a common luxury; sandals have given place to leathern boots and shoes; warmer clothing is worn in place of the light cotton garments formerly in vogue; and ornaments are more largely used by women. Clerical establishments are all recruited from India. Their

[1] One *seer* is equivalent to about 2 lb.

standard of living is somewhat high and leaves little opportunity for saving. A middle-class clerk generally has a house with two or three rooms, a kitchen and bath-room. His furniture consists of two or three chairs, a small table, lamps, carpets, rugs, and cooking utensils. He generally has one servant who is his cook and does other household work. He has two meals a day, morning and evening.

There is a considerable difference between the mode of living of a headman owning land in a village and of an ordinary cultivator. The former generally wears clothes of a superior quality, and he adds to the number a thick coat and waistcoat. He has larger house accommodation and more furniture, and he possesses a sufficiently large number of cooking utensils, rugs, blankets, felts, quilts, and saddle-bags. Both utensils and bedding in an ordinary cultivator's house are scarce. One or two earthen or metal pots, two or three bowls, an iron tripod, and a few ragged quilts complete his equipment. His dress in summer costs about Rs. 6 : a turban at Rs. 1–8–0, and a shirt, trousers, shoes, and sheet at about R. 1 each. In winter he adds a felt overcoat costing Rs. 3 and sometimes a waistcoat at Rs. 2. His wife's dress, which consists of a wrapper, a shift, wide drawers, and shoes, costs about Rs. 4.

Generally speaking, the country is scantily clothed with vege- *Reserved* tation, trees are few in number, and most of the hills which *forests.* are not protected by other and higher ranges are bare of forest growth. In Administered territory steps were taken in 1880 to control certain forest areas in the Sibi District, and rules were issued for their management in 1881. Legal and systematic action commenced in 1890, when the Forest Law and Forest Regulation Acts were enacted. Since then twenty-seven tracts have been 'reserved,' comprising a total area of 203 square miles. Some tracts in Zhob, which have been hitherto protected by the Political Agent, are now being brought under the Forest department. Between 1891 and 1900 the forest revenue averaged Rs. 17,102 and the expenditure Rs. 37,531, leaving an annual deficit of Rs. 20,429. The revenue in 1900–1 was Rs. 16,927 and the expenditure Rs. 29,254. Owing to a change of system the deficit has now been reduced, and in 1902–3 the revenue was Rs. 19,336 and the expenditure Rs. 23,240. These figures exclude the revenue and expenditure of the Zhob forests, the income of which was Rs. 6,370 in 1901–2 and the expenditure Rs. 1,713. The greater part of the revenue is derived from the sale of timber and fuel, the annual income from this source averaging nearly Rs. 15,000. The Reserves are of three kinds :

juniper forests, pistachio forests, and mixed forests. The first of these, bearing *Juniperus excelsa*, form the principal Government forests in Baluchistān. They are twelve in number, covering an area of 114 square miles. The two Reserves which contain *Pistacia khanjak* have an area of 13 square miles. Mixed forests number eleven, with an area of 78 square miles. The principal trees in these forests are *Prosopis spicigera*, *Capparis aphylla*, *Tamarix indica*, *Tamarix articulata*, *Dalbergia Sissoo*, *Olea cuspidata*, *Pistacia khanjak*, and *Acacia modesta*. In Zhob *Pinus Gerardiana*, *Pinus excelsa*, and *Olea cuspidata* are the commonest trees in protected areas. The regeneration of juniper and pistachio has not been very encouraging, except in areas closed to grazing. Here also the improvement in undergrowth indicates the benefits to be derived from the exclusion of browsing animals, especially goats and camels. Experiments in the introduction of exotic trees have not been successful, except in the case of fruit and one or two roadside trees.

No protected and legally recognized unclassed forests exist, but certain trees growing on land at the disposal of Government have been 'reserved' and their cutting is regulated. These include, besides those mentioned above, *Pistacia mutica*, *Fraxinus xanthoxyloides*, *Zizyphus nummularia*, *Zizyphus oxyphylla*, *Tecoma undulata*, *Populus euphratica*, and *Periploca aphylla*.

Forest produce. Pine timber is used for building purposes at Fort Sandeman, and juniper at Ziārat. In the rural villages, almond, apricot, mulberry, and *sinjid* wood (*Elaeagnus angustifolia*) are used for roofing. Minor forest produce includes the gum of the wild almond, cumin seed, hyssop, juniper berries (which are used for flavouring tobacco in Calcutta and Kanauj), the seeds of the edible pine, and the wild pistachio. The latter is much prized as an article of food by the natives. Asafoetida (*hing*), the gum of *Ferula persica*, is found in parts. In the lower tracts of Kalāt and Las Bela dwarf-palm, gum arabic, bdellium, honey, and shellac occur.

Administrative. The Forest department is in charge of an Extra Assistant Conservator of Forests, borne on the Punjab Provincial list, who is styled the Chief Forest Officer, and who works under the general control of the Revenue Commissioner in forest matters. The Reserves are divided into three ranges, known respectively as the Quetta, Ziārat, and Sibi range. Each range is in charge of a Deputy-Ranger, who is assisted by forest guards. The extraction of timber and fuel is carried on by unregulated fellings. The sale of minor forest produce, such as grass, fruit, and flowers, &c., is conducted by public

auction or by permit.   The relations of the Forest department
with the people of the country have always been conciliatory.
Minor forest offences average 208 per annum.

Timber for building purposes and fuel is imported from Sind. Fuel
The question of fuel supply was considered at a conference in supply.
1891, and it was decided that the main object of Government
should be to maintain existing and future forest Reserves intact
for use in times of emergency.   Government departments
within reach of the railway are, therefore, supplied from
external sources and special railway rates are allowed.   The
average area of 'reserved' forest closed to grazing is 126 square
miles.   In the other parts of closed areas grazing by right-
holders is permitted.   All these areas are available for relief
in times of scarcity.   The question of the depletion of the
undergrowth in the large grazing tracts near the towns is one
of some difficulty.

Coal is the only mineral produced in large quantities. Mines and
Petroleum has also been worked, and a syndicate has recently minerals.
been formed for extracting chromite in the Quetta-Pishīn District.   The production of coal has been 122 tons (1886), 10,368 Out-turn.
tons (1891), 24,656 tons (1901), and 47,374 tons (1903) ; and of
petroleum 27,700 gallons (1886), and 40,465 gallons (1891).   In
1903 the output of chromite amounted to 284 tons.   Earth-salt
is manufactured chiefly in Kachhi and along the coast.   It is
also obtained in Quetta-Pishīn and part of Zhob, and from the
Wad-i-Sultān in the Hāmūn-i-Māshkel.   A salt-mine is worked
in Las Bela.   The average annual out-turn of earth-salt is
estimated at about 1,000 tons.   Lime is burnt at Quetta and
also in Las Bela.   Saltpetre is manufactured in small quantities
in Kachhi.

Coal is fairly widely distributed in the Central Brāhui range, Coal.
and is worked at Khost, in the Sor range near Quetta, and in
the Bolān Pass.   It generally possesses good steam-producing
qualities, but is very friable.   The seams vary from 6 inches to
4 feet in thickness, outcropping in hill-sides and dipping steeply,
and are worked by excavating adits horizontally from the face
of the hills.   The principal colliery is at Khost, where the mines
are worked by the North-Western Railway under European
supervision.   Capital to the total amount of about $3\frac{1}{2}$ lakhs
has been invested in the undertaking.   The number of men
employed daily is about 700.   A miner earns about 12 annas
a day.   The working cost has recently been reduced to about
Rs. 8 per ton.   The miners are chiefly Makrānis, but Hazāras
and local Afghāns are also employed.   The output, which

amounted to a total of 246,426 tons between 1887 and the end of 1903, is almost entirely consumed by the North-Western Railway. In the Sor range, in which coal is to be traced for some seventeen miles, and in the Bolān Pass leases have been granted of small stretches of coal to five lessees, Government finding the supervising staff necessary for periodical inspection and to ensure safe and scientific working. The coal has to be transported by camels, and is all consumed in Quetta. About 100 men are employed daily.

Petroleum. The presence of petroleum is indicated at Shorān in Kachhi, in the Bolān Pass, in the Harnai valley, and at Khattan in the Marri country. A boring was made at Kirta in 1889 and a show of oil was struck at 360 feet, but it was afterwards abandoned, as also was one at Spīntangi. The borings so far undertaken have been made in far less promising strata than the Siwāliks, which have not been tested ; and there is no prima facie reason why mineral oil should not be discovered in the natural reservoirs of this geological group, which has produced it in Burma and Persia. Operations were carried on by Government at Khattan for seven years from 1884 to 1892, and by a private company in 1893-4, but both ventures were ultimately abandoned. Thirteen bore-holes were put down, the deepest being 736 feet, but oil was not obtained below 332 feet. It was pumped to the surface. On analysis the oil was described as containing 45 to 55 per cent. of pitch, with 45 to 35 per cent. of lubricating oil, but no light oils whatever. The total output between 1886 and 1892 was 777,225 gallons, the largest annual amount being 309,990 gallons in 1889. The private company afterwards extracted 60,000 gallons of oil. The expenditure incurred by Government amounted to about 6½ lakhs of rupees, and there was a net loss of about 4 lakhs ; but the oil may yet prove valuable for the manufacture of patent fuel. The area prospected lies in the territory of the Marri chief, who was paid Rs. 300 a month by Government during the operations, but he afterwards compounded with the private company for a lower sum.

Unworked minerals. Little is known of the unworked minerals of the country. Chrysotile, also known as fibrous serpentine or Canadian asbestos, occurs in some quantities in the Zhob valley and in the Quetta-Pishīn District. Samples of the fibre have been found to be of some commercial value. Experts have pronounced clays obtained from the Bolān Pass as fit for the production of good paint and terra-cotta and of fair Portland cement at remunerative rates. Oriental alabaster is obtainable in Chāgai, and copper, lead, iron ores, and alunogen or hair-salt have also been found there. Ferrous sulphate (melanterite), known locally as *zāgh*, is obtainable in the east of the Jhalawān country, and is used by the natives for dyeing purposes. Carbonate of lead (cerussite) is found at Sekrān near Khuzdār. A sulphur-mine was worked by the Afghān rulers near Sanni in Kachhi, but it has been abandoned since the British occupation. Iron bisulphide (marcasite) is of frequent occurrence, but it has nowhere been found in sufficiently large quantities to render it commercially valuable.

Existing conditions in Baluchistān are still too primitive to admit of the organization of industries on commercial lines. Such as there are consist of handicrafts worked at home, and in the majority of cases the work is done by the women in their spare time. In all instances the same worker completes the article in hand from the raw material down to the finish. All Baluchistān art-work displays specially Persian characteristics. Arts and manufactures

Cotton-weaving is a moribund industry still existing in a few parts. The cloth, known as *kora* in the east and *bīst dasti* in the west, is woven in pieces about 30 feet long by 28 inches broad. Coloured double sheetings called *khes*, which are fashionable among the natives, are also manufactured. Silk-weaving is done in Makrān alone. The best specimens are tartans, known as *man-o-bas*, and a dark-green crape with crimson border called *gūshān*. They resemble fabrics made at Purnea and Chittagong. Cotton- and silk-weaving.

Embroidery is very common, especially among the Brāhuis. It is highly artistic and of many varieties, but unfortunately the products have been injuriously affected by the introduction of aniline-dyed silks. Of the Brāhui embroideries, that called *mosam* is the best. It consists of very close work in a form of satin stitch, the design being primarily geometric. Other kinds, which are not quite so fine, are known as *parāwez* and *pariwār*. These embroideries are generally made in four pieces : a pair of cuffs, a breast-piece resembling the linen front of a European shirt, and a long panel forming the pocket. Another fine kind is the Kandahār embroidery, which is generally worked with a double satin stitch in ivory white. Padded or quilted embroidery is also not uncommon. In the Marri and Bugti hills the prevailing designs consist of medallions, made up of zones of herring-bone stitch separated by rings of chain stitch. In Las Bela a fine embroidery is done on silk and cloth with the crochet needle. In Kachhi, Kalāt, Las Bela, and Makrān Embroidery.

shoes, sword-belts, and other leathern goods embroidered in silk are popular. The Kachhi embroidery is exceedingly elaborate.

Carpets and rugs.

The articles known as Baloch rugs are not an indigenous product of Baluchistān. They are chiefly made at Adraskan, a place south of Herat, and in Seistān. Large quantities are, however, imported through Quetta. Saddle-bags and nose-bags made in this style are popular among Europeans for cushion covers, chair backs, &c. A few pile carpets are made in the Jhalawān country, but entirely for home use. Rugs in the *darī* stitch are manufactured in almost every nomad household. They are made for sale in some quantities by the Angārias of Las Bela and in the Bārkhān *tahsīl* of the Loralai District. Saddle-bags and nose-bags richly ornamented with shells are also made there. The *shufi*, or long rug usually stretched in front of the bedding in a nomad tent, is manufactured in Khārān and the Sarawān country. Felts of excellent quality and richly embroidered are also made, but chiefly for home use. Among minor woollen products, manufactured chiefly from camels' and goats' hair, are ropes and grain-bags, blanketing for tents, girths, and camel gear.

Factory industries.

Since the British occupation four steam flour mills have been opened. There are also two ice factories and a steam press for chopped straw, wool, and oil. Patent fuel is manufactured from coal-dust at Khost. A brewery has been started near Quetta, the out-turn of which has risen in the eighteen years between 1886 and 1903 from 86,000 to 347,000 gallons. A plentiful supply of unskilled labour is available for these industries, chiefly recruited from trans-border Afghāns and Makrānis.

Minor industries.

Among minor industries may be mentioned tanning, the manufacture of carbonate of soda, mat and basket-making, and indigenous methods of dyeing. Tanning is in vogue chiefly in Kachhi, Las Bela, and Makrān. A good soft leather is produced by immersing the hides in lime and carbonate of soda, and afterwards tanning them with a decoction of the exudation of the tamarisk. The manufacture of carbonate of soda, chiefly from the salt-worts known as *Haloxylon Griffithii* and *Haloxylon salicornicum*, is increasing. The white variety is preferred to the black. The salt-worts are cut and after being partially dried are set on fire. Much matting and raw material for mat-making is exported from Baluchistān, especially from the lower highlands and Makrān. For this purpose the dwarf-palm (*Nannorhops Ritchieana*, called *pish* or *dhorā* in the vernacular) is employed. In 1900-1 the exports of mats and raw material to Sind were valued at Rs. 44,800. The people are well versed in the manufacture of natural dyes from lac, decoctions of willow, and olive leaves, madder, &c. Pomegranate husks, alunogen or hair-salt, and lime are used as mordants. In Quetta rose-water and *attar* of roses are manufactured by Punjab Khojas from the common Persian rose. Experiments recently made in sericulture have proved successful, and Quetta silk has been pronounced of the best quality. Measures are now being taken to develop the industry.

The indigenous trade of the Province was insignificant in former times. The country owed such commercial importance Former trade. as it possessed to its geographical position athwart the main lines of communication between Persia, Central Asia, and India. The routes followed by caravans lay through the Gomal Pass to Multān, through the Harnai, Bolān, and Mūlā Passes to Shikārpur, and via Kalāt and Bela to Sonmiāni; but trade was greatly hampered by the raiding proclivities of the adjacent tribes and the system of levying transit-dues. In the earliest engagement between the British and Kalāt an attempt was made to regulate the latter, but without much success. The levy of transit-dues still constitutes one of the greatest impediments to trade in the Kalāt State. In Administered areas the system has been broken down, generally by the expropriation of right-holders. The general character of the trade between India and Baluchistān in pre-British days resembled the land trade now carried on with Afghānistān, exports from Baluchistān consisting of wool, dried fruits, medicinal drugs, and horses, and imports of metals, piece-goods, sugar, and indigo. The traders were chiefly Bābi Afghāns, Hindus of Sind, whose transactions extended far into Central Asia, and Powindah Afghāns.

Excluding internal trade, the commerce of Baluchistān divides itself naturally into two classes—trade borne by sea, Existing trade. land, and rail to and from other Provinces in India; and foreign maritime and land trade. Omitting land-borne trade with Indian Provinces, the total trade was valued in 1902-3 at more than 2 crores of rupees, a striking evidence of the prosperity engendered by British rule. Exports consist chiefly of wool, dried and fresh fruits, medicinal drugs, fish and shark-fins, raw cotton, and mats; imports, of piece-goods, food and other grains, metals, and sugar. The chief maritime centres of trade are Gwādar and Pasni; inland marts are Quetta, Sibi, Nushki, Kila Abdullah, Bhāg, and Gandāva. Much trade

also finds its way direct to large markets in Sind, such as
Jacobābād, Shikārpur, and Karāchi.

Internal trade. All goods moving within the country, otherwise than by
rail, are carried by camel. They consist chiefly of wool,
agricultural produce, and fruit, including dates. Trade is
almost entirely conducted by Hindus from India, but there
are a few Muhammadan traders from Kandahār, and along
the coast Khojas are fairly numerous. The Hindus move
with the Brāhuis, up in summer and down in winter, collecting produce at convenient centres whence they send it direct
to the larger markets. Barter is common, food-grains being
exchanged for salt, fish for dates, and cloth for *ghī* and wool.

	By Sea.			By Rail.	
	1890–1.	1900–1.	1902–3.	1900–1 *.	1902–3.
IMPORTS.	Rs.	Rs.	Rs.	Rs.	Rs.
Cotton—					
(a) Piece-goods	2,57,366	2,95,475	2,32,133	29,84,085	26,87,840
(b) Other articles	5,979	7,222	1,667	28,564	1,01,454
Grain and pulse	2,97,631	1,81,433	3,21,389	7,74,424	18,09,222
Provisions	145	1,919	815	5,40,891	6,43,812
Metals and manu-factures of metal	...	...	...	2,48,890	7,27,740
Sugar	...	8,099	15,669	8,12,034	8,31,087
Woollen piece-goods	...	...	...	81,536	1,43,248
All other articles	1,35,763	96,141	95,190	44,43,352	57,88,519
Total	6,96,884	5,90,289	6,66,863	99,13,776	1,27,32,922
EXPORTS.	Rs.	Rs.	Rs.	Rs.	Rs.
Animals, living—					
(a) Horses, ponies, mules, &c.	...	...	...	1,09,710	1,90,050
(b) Others	...	31	...	...	81,484
Cotton, raw	69,749	74,075	1,66,279	...	...
Drugs, medicinal	...	...	...	5,49,750	7,32,780
Fruits	27,105	21,023	20,661	7,72,902	7,19,400
Grain and pulse	3,702	45,665	40	1,05,568	2,49,021
Provisions—					
(a) Fish, including shark-fins	2,13,072	2,46,336	2,06,142	...	...
(b) Ghi	54,205	1,37,911	584	3,61,120	10,974
(c) Other kinds	216	1,556	...	...	...
Wool, raw	1,21,595	96,923	96,320	22,30,095	14,66,016
All other articles	3,71,745	1,59,688	2,04,877	23,82,643	41,98,364
Total	8,61,389	7,83,208	6,94,903	65,11,788	76,48,089

* The first year for which figures are available.

External trade with India. The maritime trade is carried on from the coast of Makrān
chiefly with Bombay and Karāchi, salt-fish, shark-fins, mats,

wool, and raw cotton being exchanged for cotton piece-goods, food-grains (chiefly rice), and also sugar. Trade by land passes across the frontier to the Frontier Province, the Punjab, and Sind, but it is not fully registered. Wool, *ghī*, dwarf-palm for mat-making, and sheep and goats are the chief exports, and piece-goods and food-grains the imports. Of trade carried by rail, raw wool, fruits, medicinal drugs, and *ghī* constitute the largest articles of export; piece-goods, grain, metals, and sugar are the largest imports. The table on the opposite page exhibits the value of the maritime and rail-borne trade with other parts of India.

No statistics of foreign trade by sea are available; but Foreign native craft carry dates and matting from Makrān to the trade. Persian Gulf, Arabia, and the east coast of Africa. Trade by land is carried on with Persia and Afghānistān, but is much hampered by the fiscal policy of those countries. The export of wheat, *ghī*, and horses from Afghān territory is entirely prohibited, though a good deal of smuggling takes place. The export of almonds is a State monopoly, while imports are liable to heavy duties. Transport too is entirely on camels and donkeys. The trade is registered at Chaman and Nushki, but, as goods cross the frontier at many other points on both the north and the west, the statistics are far from complete. Trade along the newly opened route from Nushki to Persia has to face keen competition from Russian goods brought from the Trans-Caspian Railway through Meshed, but, in spite of these drawbacks, it exhibits a considerable expansion. The table on the next page shows the value of the chief imports and exports of foreign land trade. The decrease in 1902–3 was due to a long period of drought which culminated in 1900–1.

The impression created on the wild tribes of the frontier by Means of the construction of railways and roads has been immense; and communica-tion. their civilizing influence has been felt far beyond the political frontier, for every year many thousands of trans-border Afghāns travel to India by their means to find remunerative employment. All the railways and the best of the roads have had their origin in strategical needs. The necessity of a railway was forcibly demonstrated by the waste of treasure and life which occurred during the second Afghān War in the weary marches between Sind and Quetta, and the Sind-Pishīn Railway owed its inception to this period. The critical year of 1885 caused the extension of the railway up the Bolān Pass, and shortly afterwards the Bolān road was bridged and metalled. The Pishīn-Dera Ghāzi

Khān road was constructed to control the country between Pishīn and the Punjab, and to form an alternative line of communication with the Indus valley.

	1897–8.*	1900–1.	1902–3.
	Rs.	Rs.	Rs.
IMPORTS.			
Animals, living	1,37,214	3,00,107	2,02,670
Drugs and medicines . . .	75,853	85,522	1,64,011
Fruits	4,50,095	3,89,256	2,40,038
Grain and pulse	2,18,312	91,548	17,484
Provisions—			
(a) Ghī	1,15,897	4,75,412	90,778
(b) Other kinds . . .	24,014	7,265	640
Wool, raw	2,08,536	2,62,125	1,71,588
All other articles	39,084	2,23,299	2,23,646
Total	12,69,005	18,34,534	11,10,855
	Rs.	Rs.	Rs.
EXPORTS.			
Cotton piece-goods, &c. . . .	3,11,742	8,06,997	5,07,200
Dyeing materials	55,043	2,62,026	52,527
Sugar	6,217	42,575	47,878
Tea	25	51,585	24,716
All other articles	69,536	2,17,657	2,24,016
Total	4,42,563	13,80,840	8,56,337

* The first year for which figures are available.

Railways. The total length of railways open for traffic was 277 miles in 1891, which had increased to 399 miles in 1901. The opening of the Nushki Railway, which was completed in 1905, has brought the total up to 481 miles. The railways, which are on the standard gauge, have been built at immense cost. The total capital outlay, excluding the Nushki Railway, has been 11 crores, equivalent to 7·3 millions sterling, or an average cost of nearly 3 lakhs per mile. The Sind-Pishīn line enters the Province near Jhatpat, traverses the Harnai valley, and has its terminus at Chaman. Its total length, including the branch line from Bostān to Quetta, is 312 miles. It was begun in 1879, but not completed to Chaman till 1892. The chief difficulties in construction were met with in the Chappar Rift, at Mud Gorge, and in excavating the KHOJAK tunnel. The Chappar Rift is a huge gorge traversing the Chappar mountain at right angles to its general strike. It is crossed by a single span bridge, 150 feet long and 234 feet high, which is reached by tunnels excavated in the solid rock with the aid of adits horizontal to the face of the cliffs. At Mud Gorge slips used to obliterate the line entirely at times, but, since 1894, the construction of

a system of dams has prevented further subsidence. The surface line originally constructed from Rindli through the Bolān
Pass was re-aligned in consequence of serious damage done by
heavy floods in 1889 and 1890, and was taken from the Nāri
river to Mushkāf and thence along the east of the valley. The
tunnelling on this line, especially between Mushkāf and Pishi,
was extremely heavy. The gradients are also steeper than those
on the Sind-Pishīn line, and are as much as one in twenty-five
between Mach and Kolepur. The length of the Quetta-Nushki
Railway is 82 miles. The principal works are the Nishpa tunnel,
half a mile long, and the excavation of the alignment through
the Shaikh Wāsil gorge. The cost will probably be less than
1 lakh per mile.

There has been a continual increase of roads since the British Roads.
occupation, connecting remote parts with the railways and the
Punjab. The principal extension has taken place in the north-
east corner of the Province. In the south and the west no
cart-roads exist, and many of the routes are barely practicable
even for camels. The following table gives statistics of the
mileage and character of the roads maintained under the supervision of officers of the Military Works service, who are in
charge of civil works in the Province :—

	1901. Mileage*.	1903. Mileage.
Cart-roads bridged and metalled † .	115	120
Cart-roads partially bridged and metalled	667	668
(a) Maintained from military funds	420	421
(b) Maintained from civil funds .	247	247
Tracks and paths .	1,602	1,628
(a) Maintained from military funds	128	134
(b) Maintained from civil funds	1,474	1,494
Total roads and paths	2,384	2,416

* In 1891 the total mileage of roads and paths was 1,520. Details are not available.
† These roads are maintained from military funds.

The total cost of maintenance of these roads in 1902–3 was
Rs. 74,919. In addition, 1,128 miles of roads and paths are
supervised by civil officers, of which 1,101 miles are maintained from Provincial revenues and 27 from Local funds.
In directly Administered areas the more important roads are
the Bolān-Quetta-Chaman road, the Pishīn-Dera Ghāzi Khān
road, and the Harnai-Fort Sandeman road. A road partially
bridged and metalled runs from Quetta through Kach and
Ziārat to Smāllan on the Harnai-Fort Sandeman road. All
these roads are provided with rest-houses or dāk-bungalows at

convenient stages. The roads from Sibi to Quetta, from Dera Ghāzi Khān to Pishīn, and from Harnai to Loralai are maintained from military funds. The maintenance of the remainder, with some minor exceptions, falls on Provincial revenues, Local funds being applied only to the maintenance of roads in the Quetta municipality and a few other headquarters stations.

Of the more important roads, that in the Bolān Pass was commenced in 1886–8, and completed to Quetta in 1888, at a total cost of about 19 lakhs. The section between Quetta and Chaman (distance 78 miles) was improved and completed between 1887 and 1893. The Pishīn-Dera Ghāzi Khān road was commenced in 1886 and completed in 1888, at a total cost of 7 lakhs. The length in Baluchistān is 224½ miles. The section of the Harnai-Fort Sandeman road between the former place and Loralai was constructed as a military road after the occupation of the Bori valley in 1886, and it cost Rs. 10,600 per mile. It traverses the fine Dilkūna gorge, which has been negotiated by carrying the road along the cliffs above flood-level. The road has been extended by civil agency from Loralai to Fort Sandeman. An important link of communication between Zhob and the Punjab will shortly be secured by the road through the Dhāna Sar in the Sulaimān range to Dera Ismail Khān, which is now being re-aligned and reconstructed at great cost. It is 115 miles long, of which about 47½ miles lie within Baluchistān. The remaining fair-weather paths and tracks form a network connecting all the principal places in Administered areas, but they are, as a rule, only fit for camel carriage. On the west the Nushki-Seistān trade-route, 378 miles to Robāt Kila, has been aligned at a cost of Rs. 29,864. Camel carriage is everywhere the ordinary means of transport, but donkeys are largely employed for light loads. In Kachhi use is made of bullock carts of the type in vogue in Sind.

Steamers. The steamers of the British India Steam Navigation Company carry passengers, mails, and cargo between India and Pasni and Gwādar on the coast. These ports are visited on alternate weeks. Owing to shoal-water, a landing can only be effected in country boats.

Post office. The postal service is under the Deputy-Postmaster-General of the Sind and Baluchistān Circle. In 1881 there were 19 post offices and 453 miles of postal line, in addition to the railway. In 1902–3, 54 post offices and 27 letter-boxes were open, and the miles of communication numbered

2,281; the letters dealt with numbered 1,201,580; post-cards, 811,030; packets, 150,745; newspapers, 208,050; and parcels, 27,740.[1] The total amount of savings bank deposits was 7·8 lakhs. The money-order system is generally utilized by the Indian population temporarily resident in the country, and also by Afghān merchants trading with India. Beyond the railway, mails are carried by horsemen, who are appointed by the District officers, and whose pay forms a Provincial charge. The levies so employed numbered 214 in 1903–4. A postal service in Las Bela has been organized by that State between Karāchi and Bela.

The first telegraph line constructed in Baluchistān was the Tele-Indo-European system, which reached Gwādar in 1863. Treaties and arrangements are in force with Las Bela and Kalāt for the protection of the line. It skirts the coast for 399 miles. The rest of the country, especially the north-east, is well provided with telegraphs, and a line runs to Robāt Kila on the Persian frontier, which also connects with the Indo-European system. The tribesmen through whose country the lines pass are responsible for their protection, with the exception of the line to Kalāt, for which a small establishment is maintained. *graphs.*

Actual famine is unknown, but scarcity is frequent. Culti- *Famine.* vation depends either on permanent or on flood irrigation, and *General conditions* as *kārez* are most numerous in the upper highlands, these *and famine* areas are better protected than the lower parts. Everywhere, *warnings.* except in the plains, the principal harvest is reaped in the spring, the chief crop being wheat. In the plains, *jowār* is the staple food-grain. Trade returns show that the average aggregate imports of food-grains by rail exceed the exports by about one-third; but much of the imported grain must be consumed by the foreign residents, and a fair wheat harvest is probably sufficient to carry the native-born population through the year. Again, the majority of the people are both graziers and agriculturists, and though the season may be unfavourable to agriculture, it may still be one of fairly good pasturage. Only a combined failure of both crops and grazing for consecutive seasons causes a crisis. Recent experience indicates that, while the people can tide over two years of bad rainfall or snowfall, a third reduces them to straits. Prices rise and a large emigration takes place to more favoured tracts.

Local tradition speaks of constant scarcity, and Masson *Historical.*

---

[1] The figures for post offices and miles of communication do not include 372 miles of *dāk* line from Nushki to Robāt Kila and the four post offices located thereon.

records a ten years' drought from 1830 to 1840. Between 1897 and 1901 a succession of bad years resulted in a deficit of land revenue of about $1\frac{1}{2}$ lakhs of rupees, an expenditure of Rs. 1,87,443 from special Imperial grants on works and relief, and of Rs. 30,000 from the Indian Famine Relief Fund. Large advances were also made for agricultural purposes. Produce revenue adjusts itself automatically; and during the first two years some remissions and suspensions in assessed areas, combined with assistance for the purchase of seed and stock, were found to be all that was required. But, on the culmination of the drought in 1900-1, relief works had to be opened, chiefly roads, and doles of grain were distributed to the Marri and Bugti tribes. Advances and doles to the amount of about Rs. 34,000 were also made by Native States. No mortality was recorded.

Protection.   The greatest safeguard against famine consists in the migratory habits of the people, and the proximity of fully protected areas in Sind, where ample means of subsistence exist for all who are willing to work. The two state irrigation schemes in the upper highlands are dependent on rainfall, and cannot, therefore, be regarded as entirely protective. Except in Kachhi it is doubtful if there is much scope for other large schemes. The widest source of protection probably lies in the extension of embankments for catching the rain-water as it runs off the stony sides of the hills. There are indications that large resort was had to this method of retaining the moisture in prehistoric times, the *gabrbands* of the Jhalawān country having been undoubtedly intended for this purpose.

Administration.   The head of the local administration is the officer styled Agent to the Governor-General and Chief Commissioner. The following is a list of those who have held the substantive appointment: Colonel Sir Robert Sandeman (1877), Major-General Sir James Browne (1892), Mr. (now Sir Hugh) Barnes (1896), Colonel C. E. Yate (1900). The Agent to the Governor-General exercises judicial powers under the Frontier Crimes Regulation, and conducts the political administration of Baluchistān. He is also Inspector-General of Police and Levies. He has two Assistants, who are officers of the Political Department, and a personal Native Assistant of the rank of an Extra Assistant Commissioner. Other members of his staff are the Agency Surgeon and the Officer Commanding the Royal Engineers at Quetta, who is the Civil Secretary in the Public Works department.

Next in rank comes the Revenue Commissioner, who advises the Agent to the Governor-General in financial matters and generally controls the revenue administration. He combines the functions of Settlement Commissioner, Superintendent of Stamps, Commissioner of Excise, Inspector-General of Registration and Jails, Registrar-General, and Judicial Commissioner ; and he has the powers of a Local Government in the disposition of Local funds. Forest administration is also in his hands. For local and rural administration the Political Agency or District is the unit. Each District is divided into *tahsīls*, of which one or more constitute a subdivision. The village, which is often nothing more than the area occupied by a tribal subdivision, is the unit of management within the *tahsīl*. The Political Agent, who is also Deputy-Commissioner for such portions of his District as form part of British India, is the Collector, District and Sessions Judge, and the administrative head of his charge. He has frequently to deal with important trans-border affairs of a political character. Assistant Political Agents and Extra Assistant Commissioners or Native Assistants are in charge of subdivisions, supervise the collection of revenue, exercise civil and criminal powers, and have the subordinate political control of the tribes in their respective areas. Each *tahsīl* is in charge of a *tahsīldār*, with an assistant styled *naib-tahsīldār*. A *naib-tahsīldār* holds charge of each sub-*tahsīl*. These officials are primarily responsible for the collection of the revenue, but they also exercise judicial powers. A *tahsīl* is distributed into circles, in which the *patwāri* represents the authority of Government. He is charged with the maintenance of settlement, crop and other records, supervises the maintenance of sources of irrigation, and assists in maintaining order. It is his duty to see that the headmen collect and pay the revenue demand punctually. The village headmen collect the revenue, assist in the appraisement of crops, and maintain order. They are generally remunerated by an allowance of 5 per cent. on the gross collections. The Sandeman system of offering allowances to headmen to maintain followers, by whose means they are expected to control their tribes, is freely employed. It is closely connected with the system of levy services referred to below. In 1904 five Political Agents, seven Assistant Political Agents, eight Extra Assistant Commissioners, and five Native Assistants were employed in District administration in the Province. There were also fourteen *tahsīldārs* and eighteen *naib-tahsīldārs*. A Cantonment Magistrate and an Assistant Cantonment Magistrate are posted to Quetta, and a Staff Officer performs the same duties in Loralai. A Munsif is stationed at Quetta and another at Sibi. The Director, Persian Gulf Telegraphs, is a Justice of the Peace for places along the coast, and also exercises certain political powers as an Assistant to the Political Agent in Kalāt.

Adminis-
trative
divisions.
Excluding Native States, the Province is divided into two main portions: British Baluchistān, and the territories administered by the Agent to the Governor-General, the latter being generally known as the Agency Territories. British Baluchistān includes the *tahsīls* of Shāhrig, Sibi, Duki, Pishīn with Shorarūd, and the Chaman subdivision, with a total area of 9,476 square miles. Agency Territories is an elastic term, including areas which are directly administered and also other tracts which are merely politically controlled. Of these, the directly administered areas cover 37,216 square miles, comprising the Quetta *tahsīl*, the Bolān Pass District, and the Nushki and Nasīrābād *tahsīls*, all of which have been leased from the Khān of Kalāt; the lands occupied by the railway from Jhatpat to Mithri, Nāri to Spīntangi, and Spezand to Sorosham; the Chāgai and Western Sinjrāni country; and the whole of the Zhob and Loralai Agencies, except the Duki *tahsīl* in the latter. The inhabitants of the last two tracts have placed themselves under British protection from time to time. Throughout these areas revenue is collected. The part of Baluchistān, therefore, including British Baluchistān, which is under direct administration, covers 46,692 square miles, an area about the same size as Sind. It comprises six Districts—Quetta-Pishīn, Sibi, Loralai, Zhob, Chāgai, and the Bolān Pass. Each of the first four is in charge of a Political Agent and the fifth of a Political Assistant, while the Bolān Pass is administered by the Political Agent in Kalāt. The latter, with two Political Assistants, one of whom is *ex officio* Commandant of the Makrān Levy Corps, controls the affairs of the Kalāt and Las Bela States. The Political Agent in Sibi exercises political control over the Marri and Bugti tribes and in the Lahri *niābat* of the Kalāt State, which is inhabited by the Dombki, Umrāni, and Kaheri tribes.

The Kalāt
constitu-
tion.
The Native States are two in number, KALĀT and LAS BELA, the latter being nominally a feudatory of the former. The Kalāt State consists of a confederacy of tribal groups headed by the Khān of Kalāt. These groups were originally organized into three great divisions: (1) the Khān's *ulus*, or following, which was scattered throughout the country; (2) the Sarawān tribesmen living to the north of Kalāt under their hereditary chief, the Raisāni Sardār; and (3) the Jhalawān tribesmen

living to the south of Kalāt under the Zahri Sardār. All were
liable to the Khān for military service. Succession to the
masnad of Kalāt appears to have been by inheritance, subject
to the approval of the chiefs and of the paramount power.
Gross incompetence might exclude. In external affairs the
Khān was supreme and absolute. Internally each of the
Sarawān and Jhalawān tribes retained the fullest rights of self-
government, but by the unwritten rule of the constitution there
was a general right of interference by the Khān. The Khān,
through his *naibs* or deputy-governors, managed the affairs of
those people from whom he collected revenue. Khārān was,
and still is, quasi-independent. These arrangements have,
however, been modified by lapse of time.

The relations of Kalāt with the British Government are
governed by two treaties, those of 1854 and 1876. The treaty
of 1876 reaffirmed the treaty of 1854. Under the terms of the
earlier treaty a subsidy of Rs. 50,000 was payable to the Khān,
which was raised to 1 lakh in 1876. At the same time the
Khān agreed to act in subordinate co-operation with the British
Government ; a British Agency was re-established at the Khān's
court with certain powers of arbitration ; and the presence of
British troops in Kalāt was permitted. The construction of railways and telegraphs and freedom of trade were also provided
for. There are further agreements with Kalāt in connexion
with the construction of the Indo-European Telegraph, the
cession of jurisdiction on the railways and in the Bolān Pass,
and the permanent lease of Quetta, Nushki, and Nasīrābād.
A Political Agent was permanently reappointed to Kalāt in
1884, to keep touch with the Khān and to exercise the right of
arbitration already referred to. The Khān is entitled to a salute
of nineteen guns. *(margin: Relations with the British Government.)*

The relations of the Brāhui tribesmen with the Khān are
now regulated by the Mastung agreement, the treaty of 1876,
and the custom which has arisen therefrom. In the Mastung
agreement the Sarawān and Jhalawān Sardārs declared their
submission and allegiance to the Khān, the Khān on his part
restoring to them their ancient rights and privileges and promising good treatment so long as they proved loyal and faith-
ful. Difficulties having arisen in cases where disputes had
occurred between the Khān's deputies, as representing his subjects, and the Brāhui tribesmen, such disputes were to be
referred to the Khān for inquiry and decision, and, in case of
disagreement, the disputed point was to be left to the arbitra-
tion of the British Government. The rights of internal self- *(margin: Relations of the tribesmen with the Khān.)*

government previously possessed by the tribesmen remained intact. In 1879 service and allowances were granted by the British Government to the Sarawān chiefs, in personal recognition of their loyal behaviour during the second Afghān War. To assist in the administrative control of the tribes, a Native Assistant for Sarawān was appointed in 1902.

Administrative arrangements in the Khān's jurisdiction. The Khān is assisted in the general administration of his State by a Political Adviser, whose services are lent by the British Government. The affairs of the Jhalawān country, in which certain tribal subsidies are paid by the Khān, are supervised by a Native Assistant also lent by the British Government, who is stationed at Khuzdār. The country, other than tribal, in which the Khān exercises control and collects revenue is divided into *niābats*. Makrān is under the control of an officer known as the *nāzim*. Each of the *niābats*, of which there are nine, was formerly in charge of a *naib* or deputy-governor. The Mastung *niābat* and the five *niābats* of Kachhi are supervised by four *tahsīldārs* or *mustaufis*, who are represented locally by deputies, called *jā-nashīn*. This system was introduced in 1902, but in some cases the *naib* has been retained and exercises jurisdiction concurrently with the *jā-nashīn*. Important civil cases are heard by the Political Adviser. The *mustaufis* hear civil suits up to Rs. 10,000 in value, and *naibs* and *jā-nashīns* suits of lower value on a graded scale. Court-fees are levied at 10 per cent. *ad valorem*. Criminal cases are dealt with by the Political Adviser on the basis of tribal custom.

Las Bela. The chief of Las Bela, known as the Jām, is bound by agreement with the British Government to conduct the administration of his State in accordance with the advice of the Governor-General's Agent. This control is exercised through the Political Agent in Kalāt. Sentences of death must be referred for confirmation. The Jām also employs an approved Wazīr, to whose advice he is subject, and who generally assists him in the transaction of state business. For purposes of administration the State is divided into seven *niābats*. The *naib*, or officer in charge, inquires into petty judicial cases, and collects transit-dues and the land revenue after it has been appraised by the revenue establishment. He submits all cases for the orders of the Jām. The revenue arrangements are in the hands of a *tahsīldār*, assisted by a head revenue *naib* and other *naibs*.

Legislation and justice. Acts and Regulations are extended to British Baluchistān either under the Scheduled Districts Act (XIV of 1874) or by special mention in the Act itself, and are applied to the Agency Territories by the Governor-General-in-Council under the

Indian (Foreign Jurisdiction) Order in Council, 1902. Enactments peculiar to Baluchistān are the Laws Law, the Civil Justice Law and Regulation, the Criminal Justice Law and Regulation, and the Forest Law and Regulation, all passed in 1890. The Civil and Criminal Justice Laws and Regulations were modified in 1893, and re-enacted in 1896, when the post of Judicial Commissioner was created. He is the High Court for Baluchistān, but sentences of death passed or confirmed by him require the sanction of the Local Government. In proceedings against European British subjects the Punjab Chief Court is the High Court. Each District is a Sessions division, the District Magistrate being the Sessions Judge, who may take cognizance of any offence as a court of original jurisdiction without previous commitment by a magistrate. The trial may take place without jury or assessors. Assistant Political Agents, Extra Assistant Commissioners, and Native Assistants ordinarily have the powers of a magistrate of the first class; *tahsīldārs* those of the second class, and *naib-tahsīldārs* those of the third class. The Cantonment Magistrate, Quetta, exercises first-class powers; and the Cantonment Magistrate, Loralai, and the Assistant Cantonment Magistrate, Quetta, possess second-class powers. The limit within which sentences are not appealable has been raised in cases tried by Courts of Sessions and certain magistrates.

The civil courts are of five grades : courts of *naib-tahsīldārs*, with jurisdiction up to Rs. 50; of *tahsīldārs* and Munsifs, with jurisdiction up to Rs. 300; of Assistant Political Agents, Extra Assistant Commissioners, and Native Assistants, with jurisdiction which may extend up to Rs. 10,000 ; and of Political Agents, without limit of pecuniary jurisdiction. The Judicial Commissioner constitutes the highest appellate authority. Appeals from the orders of a *naib-tahsīldār, tahsīldār*, or Munsif ordinarily lie to the subdivisional officers concerned. The Cantonment Magistrate and Assistant Cantonment Magistrate at Quetta and certain other officers preside over Courts of Small Causes. In questions relating to certain specified subjects, the civil courts are bound to have regard to tribal custom, where such custom is inconsistent with ordinary Hindu or Muhammadan law. No legal practitioner can appear in any court without the sanction of the Local Government. The number of cases disposed of is shown in the table on the next page. Nearly half the number of criminal cases are of a petty nature. The decline in the number of civil cases is due to the completion of the large railway and other works.

Jirgas.   The indigenous system of referring disputes to a council of
tribal elders or *jirga* has been developed under British administration, the Punjab Frontier Crimes Regulation[1] having been
applied for this purpose in 1890.   Local cases are referred to
local *jirgas*, while intertribal and other important cases are
decided by *Shāhi jirgas* which meet twice a year at Sibi and

	Criminal.			Civil.				Revenue and Miscellaneous.
	Total.	Appellate.	Original.	Total.	Appellate.	Original.	Execution of decree.	
Annual average for five years ending in 1897-8 .	3,700	119	3,581	10,848	258	6,767	3,823	2,397
Annual average for three years ending in 1900-1	3,917	171	3,746	8,010	199	4,741	3,070	2,555
1902-3 . .	4,161	106	4,055	7,340	174	4,385	2,781	4,064

Quetta.   A *Shāhi jirga* is also held once a year at Fort Munro,
for the disposal of inter-Provincial cases between the Punjab
and Baluchistān.   These periodical assemblies have to decide
cases of blood-feud, murder, important land disputes, &c., which,
if unadjusted, would probably lead to bloodshed, loss of life,
and political complications.   During the five years ending March,
1898, the average annual number of cases thus disposed of
was 2,871 ; in the succeeding three years the average number
was 3,049.   The actual number of cases in 1902-3 was 4,230.
The Murderous Outrages Regulation was extended to Baluchistān in 1902.   Under its provisions a fanatic guilty of murder
may be sentenced to death or transportation or imprisonment
for life in India ; to forfeiture of property ; and to whipping, in
addition to transportation or imprisonment.   It also authorizes
the enforcement of tribal and village responsibility.   During the
ten years ending 1902 the number of murderous outrages on
Europeans was nine and on natives of India sixteen.

Registra-   The Registrars under the Indian Registration Act are Dis-
tion.   trict officers, and the Sub-Registrars are generally *tahsīldārs*.
In 1893-4 the number of Registrars was 3 and of Sub-
Registrars 12 ; 632 documents were registered, registration
being compulsory in the case of 425.   In 1902-3 the number
of offices was 19 and the number of documents registered
868.

[1] The Frontier Crimes Regulation III of 1901 has since been applied
to Baluchistān, with certain modifications.

The Provincial finances are managed in accordance with Finance.
a quasi-Provincial settlement framed by the Government of
India. The first settlement was made for a period of five years
commencing in 1897; the terms were slightly modified and
renewed for a similar period in April, 1902.

Under native rule the only items of revenue, other than that Historical
derived from land, consisted of transit-dues (*sung*), fines, and,
in some places, grazing tax. The amount realized cannot be
ascertained. The transit-dues were either taken by the State,
or shared between the State and the tribal chiefs, or pocketed
by the chiefs themselves. They have now been abolished in
all Administered areas. On the first arrival of the British, the
niābat of Quetta was managed by British officers on behalf
of the Khān of Khalāt from 1877 to 1883. During this period
the annual revenue averaged Rs. 47,674. From April 1,
1883, the revenues of Quetta were treated as an Excluded
Local Fund for one year; but afterwards, up to 1891, they were
brought into the regular accounts of the Government. During
these years the revenue averaged Rs. 1,46,000 per annum. In
April, 1891, the Agent to the Governor-General was permitted
to exercise the powers of a Local Government for two years in
respect of the revenues of the Quetta District, which were
entirely made over to him. Meanwhile the provincialization
of the Police and Levies, except those in the Zhob valley, had
been authorized; the Bori, Bārkhān, and Zhob valleys had come
under control and the Zhob Levy Corps had been constituted,
and the general control of the revenue and expenditure in each
case had also been made over to the Local Government. In
1893, therefore, a consolidated contract was sanctioned which
included the four existing contracts, and an assignment of
Rs. 8,65,000 per annum for a period of four years was granted.
This was raised later on by Rs. 600 per annum. During the
four years 1893 to 1897 the receipts, excluding the Imperial
assignment, averaged Rs. 5,70,550, and the expenditure Rs.
14,20,000 annually. The revenue and expenditure of Thal-
Chotiāli—i.e. Duki, Sibi, and Shāhrig—and of Pishīn, which
areas had been declared to be British Baluchistān, had throughout this period been treated as Imperial. Their revenue from
1883 to 1890 averaged Rs. 3,40,000, and from 1890 to 1897
Rs. 3,67,000 per annum.

From April 1, 1897, the whole revenue and expenditure of First pro-
the Province, classified under certain specified heads, was vincialized
provincialized, the settlement being sanctioned for a period settlement.
of five years. The standard figure of revenue was fixed at

Rs. 9,89,000, and that of expenditure at Rs. 22,34,000. The latter was afterwards increased to Rs. 22,74,000, owing to the arrangements made in connexion with the Chāgai District. Under this settlement the Agent to the Governor-General exercises the powers usually given to Local Governments under the scheme of Provincial finance, and is subject to the rules of financial procedure which have been applied to Local Governments. The balance at the close of each year is carried on to the next. The expenditure on levies may not be materially reduced below Rs. 8,27,000 per annum without special sanction. The revenue includes that from salt made locally, usually an Imperial item ; the only head of revenue classed as Imperial is interest. Under expenditure, the items (*a*) Ecclesiastical, (*b*) Territorial and Political Pensions, and (*c*) Political Salaries, are treated as Imperial. Territorial and political pensions include sums payable to persons of political importance. The salaries of Political Officers borne on the Foreign Department list, of the Agency Surgeon and Assistant Surgeon at Kalāt, and of all Extra Assistant Commissioners and Native Assistants, except the Extra Assistant Commissioner, Quetta, and the Native Assistant, Sarawān, are debitable against the head Political Salaries.

Result of the first settlement. The ultimate result of the first provincialized settlement was a debit balance of Rs. 43,312. The average revenue amounted to only Rs. 9,36,000, about half a lakh less than the standard figure ; on the other hand the expenditure was reduced to an average of Rs. 22,39,000 per annum, against the estimated standard figure of Rs. 22,74,000. A decrease occurred under almost all heads of revenue, especially from land revenue and stamps. It was only from irrigation that an increase occurred, amounting to about Rs. 6,000. The standard figure of land revenue was Rs. 6,19,000 ; but, owing to a series of dry years, culminating in actual drought in 1900-1, the realizations averaged Rs. 5,85,000, or about Rs. 33,000 less than the estimate. Stamp revenue also fell below the estimate by about Rs. 31,000. Owing to the expansion of the Province, there have been increases in the establishments for the collection of land revenue and for general administration, and in the expenditure on levies. By the reorganization of the Forest department a saving of about Rs. 14,000 has been effected. Owing to the peculiar circumstances of the Province, the allotment for public works constitutes the only reserve from which expenditure in times of scarcity can be met. All other charges are practically fixed. During the quinquennial period

HEADS OF REVENUE.	AVERAGE FOR THE QUINQUENNIAL PERIOD ENDING 1902.		ACTUALS FOR 1900-1.		ACTUALS FOR 1902-3.	
	Total amount raised in Province.	Total amount credited to Provincial revenues.	Total amount raised in Province.	Total amount credited to Provincial revenues.	Total amount raised in Province.	Total amount credited to Provincial revenues.
	Rs.	Rs.	Rs.	Rs.	Rs.	Rs.
Land revenue . .	5,85,000	5,85,000	6,05,000	6,05,000	5,40,000	5,40,000
Stamps . . .	63,000	63,000	61,000	61,000	65,000	65,000
Excise . . .	1,37,000	1,37,000	1,51,000	1,51,000	1,27,000	1,27,000
Forests . . .	16,000	16,000	17,000	17,000	19,000	19,000
Interest . . .	5,000	...	5,000	...	9,000	...
Other sources . .	1,35,000	1,35,000	1,20,000	1,20,000	1,02,000	1,02,000
Total	9,41,000	9,36,000	9,59,000	9,54,000	8,62,000	8,53,000
Assignment from Imperial revenues .	...	12,74,000	...	12,85,000	...	16,98,000
Grand total	9,41,000	22,10,000	9,59,000	22,39,000	8,62,000	25,51,000

HEADS OF EXPENDITURE.	Average for the quinquennial period ending 1902.	Actuals for 1900-1.	Actuals for 1902-3.
	Rs.	Rs.	Rs.
Opening balance.	20,000	− 17,000	− 43,000
Charges in respect of collection (principally land revenue and forests) .	1,80,000	1,86,000	1,95,000
Salaries and expenses of Civil Departments—			
(a) General Administration .	3,36,000	3,34,000	3,52,000
(b) Law and Justice . .	71,000	77,000	70,000
(c) Police . . .	2,49,000	2,63,000	2,60,000
(d) Levies, including postal levies	8,54,000	8,75,000	8,79,000
(e) Education . . .	9,000	9,000	10,000
(f) Medical . . .	50,000	57,000	57,000
(g) Other heads (Scientific and Minor Departments) .	2,000	2,000	51,000
Pensions and Miscellaneous Civil charges (including political pensions)	9,000	7,000	16,000
Irrigation	30,000	21,000	81,000
Civil Public Works . . .	3,75,000	3,18,000	3,97,000
Other charges and adjustments, Refunds and Drawbacks . .	2,000	2,000	1,000
Assignments and compensations .	72,000	77,000	78,000
Grand total	22,39,000	22,28,000	24,47,000
Closing balance	− 9,000	− 6,000	+ 61,000

the amount spent on public works was less than the standard
by about Rs. 50,000. Of the average sum of Rs. 3,75,000
expended on public works, about one-third was devoted to
repairs and the remainder to original works. The greater part
of the assigned revenues was thus shown to be stationary,
while the expenditure on public works, and especially on
roads, was unduly curtailed. In the renewed settlement of
1902 an increase in the Provincial assignment was sanctioned,
to enable the provincialized services to be maintained in a state
of efficiency and internal communications to be developed.
The annual assignment now amounts to a total of 15 lakhs per
annum.

Revenue and expenditure. The tables on the previous page give the average revenue
and expenditure under main heads for the quinquennial period
ending March, 1902, and the actuals for the two years 1900–1
and 1902–3.

Tenures. The land tenures of the Province are of a very simple
character. In a few tracts the organized and corporate system
originating in tribal conquest and territorial allotment, often in
the shape of compensation for blood, still exists. The land
appears to have been divided originally into groups of holdings
forming the several shares of the different tribal subdivisions.
In course of time, however, successive distributions have resulted
in individual proprietorship, which is now almost universal.
Such proprietors have full control over their property and the
right of alienation. In some parts periodical division of land
and water still takes place. In others, where there is little water
in proportion to the amount of land, the water is permanently
divided, while the land is owned jointly and apportioned for
each crop. In irrigated tracts the proprietor is frequently the
cultivator. If he leases his land, it is only as a temporary
measure. In flood- and rain-crop areas, where cultivation is
only possible by embanking, it was formerly usual in the highlands for an owner to assign the proprietary right in a portion
of the land embanked, generally a half or a quarter, to the
person making the embankment. This system is falling into
desuetude in proportion as the value of land is appreciated, and
temporary written leases are now generally granted. In low-
lying tracts a cultivator who makes an embankment pays the
proprietor a fixed share in the produce (*hak-i-topa*) and acquires
an alienable occupancy right.

Revenue history. Revenue was levied in Mughal times partly in cash and partly
in kind. The tribesmen were also required to furnish a specified
number of horsemen and footmen. In 1590 Duki, Pishīn, the

Harnai valley, Shorarūd, and Quetta were included in the *sarkār* of Kandahār, and Sibi in the *sarkār* of Bhakkar. Besides revenue in cash and kind, the value of which may be estimated at about $1\frac{1}{2}$ lakhs of rupees, these areas furnished 4,750 armed horsemen and 6,400 footmen. Immediately previous to the advent of the British, various systems prevailed in different parts of the country. Tribes far remote from the head-quarters of the provincial governors paid only an occasional sheep or goat to the tribal chief, who carried a small *nazrāna* or present to the local governor. In those districts which were under more immediate control and had originally furnished armed men, this service had been commuted into a cash payment known as *gham-i-naukar*. In some places a fixed amount in cash or kind, known as *zar-i-kalang*, was levied, but a complicated system of *batai* or sharing of grain was the more general method of taking revenue. The share realized varied from one-third to one-tenth, and resort was sometimes had to appraisement. A host of cesses were also levied, including payments for the *naib* or agent, the weighman, the seal-man, and the crop-watcher. The total amount thus exacted appears to have seldom amounted to less than one-half of the produce. The *batai* system still survives in this form in the Kalāt State.

Under British rule no attempt has yet been made to enforce entire uniformity in revenue management. Existing methods have been taken as the basis for introducing an improved system, and great care has been exercised not to cause discontent among the people. The present methods of realizing land revenue are as follows : (*a*) cash assessments fixed for a term of years (*jamabast*) ; (*b*) temporary assessments (*ijāra*) ; (*c*) division of the produce (*batai*) ; (*d*) appraisement of the standing crops and levy of revenue in kind (*tashkhīs* or *dānabandi*) ; and (*e*) estimation of the Government share in cash after measurement of a portion of the crops (*tashkhīs-i-naqdi*). In the last three cases the Government share varies, the highest rate being one-third and the lowest one-eighth. The usual rate is one-sixth. Where revenue is taken in kind, an amount of fodder equal to the grain is also generally realized. As a result of inquiries into the existing revenue system made in 1882, orders were issued for making summary settlements, village by village, for the removal of irritating fees, and for the conversion in special cases of grain dues into cash. At this time the usual share of produce taken in irrigated lands was one-third ; but in 1887 this rate was reduced, and about the same time the Govern-

The British system.

ment of India recommended the introduction of fixed assessments throughout the Agency. It was found impossible to effect this change immediately, and an option was therefore given to the cultivator to pay the revenue in kind or the equivalent of the crop assessment in cash. As new areas have come under control, *batai* revenue has usually been levied at one-sixth, and the cultivator is found to prefer this system with all its drawbacks to cash payment.

Surveys and settlements. Settlement work was begun in 1902 and is still in progress. The record-of-rights has covered the *tahsīls* of Quetta, Pishīn, Hindubāgh, Kila Saifulla, Bori, Shāhrig, Sibi, and Sanjāwi. A fixed cash assessment has been introduced only in the Quetta, Pishīn, Shāhrig, and Sanjāwi *tahsīls*. This being the first settlement and the data available being scanty, the methods followed have been summary. Keeping in view the fact that only a light assessment was required, the Settlement officer fixed the revenue after personal inspection and after calculation of the average out-turn of the principal crops, the valuation of the Government share being ascertained at the average prevailing prices. Regard has also been had to the proximity of markets, the quantity of irrigated lands, the nature of the soil, and the number of crops usually obtained. In rain-crop areas the introduction of a fluctuating grain assessment, proposed by the Government of India, has not yet been found practicable ; and when a cultivable ' dry ' crop area forms even a large addition to an irrigated estate it has, as a rule, not been found worth while to assess it. Extensive tracts have been formed into separate estates, which are subject to *batai*. The period of assessment is usually ten years, but this has been extended to twenty in Pishīn. In settled *tahsīls* the incidence per acre on the irrigable area varies from about 7 annas to about Rs. 3–9–0, and on the area actually irrigated from Rs. 1–5–3 to Rs. 5. The assessments are on the whole low. They generally follow the shares in water, but are sometimes fixed on areas also. Individual holders are in all cases responsible for payment of the revenue. The ordinary term of exemption for a new source of irrigation is ten years, and land brought under cultivation during the term of the settlement is not liable to revenue. The planting of fruit-trees is also encouraged. The proprietary right in land acquired by confiscation under native rule has now passed to the British Government, and it has been granted to cultivators at rates which cover both land revenue and the proprietor's share of the produce. In pre-British days, the state was considered the owner of all grazing lands; and in the draft Land Revenue Regulation, which is now under consideration, a provision has been inserted giving the Government the presumptive right in all lands comprised in unclaimed and unoccupied waste. The Punjab Land Revenue Act, XVII of 1887, has been applied with modifications to Quetta and Pishīn, and rules have been issued for the maintenance of records. For this purpose an establishment of *kānungos*, *patwāris*, &c., is maintained.

Grazing tax (*tirni*) has been levied everywhere since 1890, except in the Bolān, the Nasīrābād *tahsīl*, and Nushki. In the latter District only nomads from Afghān territory are liable. Rates vary from one anna for a sheep or goat to R. 1–8–0 for a female camel. Plough and milch cattle are exempt. The tax is collected by annual enumeration or by annual contract without actual counting. It yielded Rs. 88,682 in 1903–4, or about 13 per cent. of the aggregate land revenue. Grazing tax.

The total land revenue in 1900–1, including Rs. 4,394 collected as grazing tax from Powindahs on their way to the Punjab, and payable to that Administration, was 6 lakhs, which gives an incidence of Rs. 1·96 per head of total population and Rs. 2·27 on the rural population. Of this, 1·7 lakhs was collected by cash assessment and 3·4 lakhs by division of crops. The annual value in 1901 of the Government revenue alienated in revenue-free grants was Rs. 88,783. Most of these grants constitute a relic of Afghān times, and are held by privileged classes such as Saiyids. Under British rule they have sometimes been made to persons for services rendered, or to chiefs to enable them to support their position or compensate them for the loss of former privileges. They consist either of an assignment of the whole or of a fixed proportion of the revenue on certain lands, of fixed allowances in grain, or of remissions of grazing tax. Remissions or suspensions of revenue are given in tracts under fixed assessment only in years of drought or damaged crops, and are based on the proportion of the area in which crops have failed. Remissions of grazing tax are also allowed. Under the Civil Justice Law agricultural land may not be sold in execution of a decree without the sanction of the Local Government, and the draft Land Revenue Regulation contains provisions limiting the power of alienation of such land to non-agriculturists. Incidence of revenue, and free grants.

The opium revenue is derived entirely from vend fees. The Opium.

cultivation of poppy is prohibited, and opium required for local consumption is imported from the Punjab by licensed vendors who make their own arrangements for procuring it. No import duty is levied. The exclusive right of retailing opium is disposed of annually by auction for each District, the number of shops being limited. The sale of opium and poppy-heads for medicinal purposes is also regulated. The consumption of opium in 1900–1 was about 16 maunds, and in 1902–3 about 15 maunds.

Salt.   The salt consumed in Baluchistān consists of earth-salt manufactured in the Province, and rock-salt. The latter is imported by rail, in small quantities only, owing to the competition of untaxed local salt. The imports of rock-salt averaged about 800 maunds during the three years ending in 1900 ; the imports in 1900–1 reached 1,142 maunds. The wholesale price at Sibi is about Rs. 3–12–0 and at Quetta Rs. 4–9–0 per maund. Punjab rock-salt is used in the towns and bazars by the non-indigenous civil and military population, while the local earth-salt is used by the tribesmen. The latter pays duty on importation into Administered areas at Rs. 1–8–0 per standard maund. Local salt on importation into Quetta town pays duty at Rs. 1–8–0, but in Pishīn bazar and Kila Abdullah the rate is R. 1. No tax is levied on earth-salt produced in Zhob, Loralai, or Chāgai. In the Native States the right of manufacture is generally given on contract. No preventive establishments are anywhere maintained. The annual revenue from salt from 1897 to 1900 averaged Rs. 3,670. In 1900–1 it was Rs. 3,383, and in 1902–3 Rs. 3,151.

Intoxicating drugs.   The cultivation of hemp has been absolutely prohibited in British Baluchistān since July, 1896. In the Kalāt highlands its cultivation appears to be on the increase. *Charas* and *bhang* are imported into British territory in small quantities from Afghānistān and Kalāt, and have been subject to import duty since 1902. The main supply of *charas* comes from the Punjab, while *bhang* and *gānja* are supplied by Sind. *Bhang* is the only drug of which there is any considerable consumption. The consumption of intoxicating drugs in 1900–1 was as follows: poppy-heads, $3\frac{1}{2}$ maunds ; *gānja*, 34 seers ; *charas*, 30 maunds ; and *bhang*, 119 maunds. In 1902–3 the consumption of poppy-heads amounted to $1\frac{1}{8}$ maunds ; of *gānja* to 8 seers ; of *charas* to $29\frac{1}{5}$ maunds ; and of *bhang* to $40\frac{3}{4}$ maunds. Separate annual contracts are given to licensed vendors for the wholesale and retail vend of intoxicating drugs. For medicinal purposes they are sold by licensed druggists. The incidence of revenue per head of population in 1900–1 was about one anna.

The manufacture and vend of country spirits are combined Liquor. under a monopoly system. Each District forms a separate farm. The rights are generally disposed of annually by auction. The sale of rum is sometimes included in the contract. Manufacture is carried on under the out-still system.

The out-turn of the brewery at Quetta averaged 212,977 Beer. gallons of beer per annum between 1891 and 1900, of which 170,460 gallons were supplied to the local Commissariat department. The quantity brewed was 238,572 gallons in 1901, and 347,220 gallons in 1903. Up to 1897 malt liquors supplied to the Commissariat department were, as a special case, free of duty. One anna per gallon is now levied.

Foreign liquors, including spirit manufactured in other parts Foreign of India after the English method, are sold under wholesale and liquors. retail licences, which are granted on payment of fixed fees of Rs. 32 for wholesale vend and from Rs. 100 to Rs. 300 for retail vend. Retail vendors may not sell by the glass. Rates varying from Rs. 6 to Rs. 200 per annum are charged for licences for places of refreshment.

The excise revenue from various sources has been as Revenue. under:—

	Average of ten years ending 1899–1900.	1900–1.	1902–3.
	Rs.	Rs.	Rs.
Country liquors, including beer and rum	1,00,980	1,05,177	94,239
Opium	17,359	17,490	12,495
Drugs	18,569	21,291	17,415
Foreign liquors	3,998	3,678	2,845
Total	1,40,906	1,47,636	1,26,994

The incidence per head is about seven annas. The consumption of intoxicating liquors and opium is confined to the foreign population. Hemp drugs are used to some extent by the people of the country. There appears to be a tendency among the educated classes to consume foreign liquors in preference to country spirits.

The net stamp revenue during the nine years 1891 to 1900 Stamps. averaged Rs. 76,600. In 1900–1 it was Rs. 59,000, and in

1902–3 Rs. 62,600. About two-thirds of the total is derived from judicial and one-third from non-judicial stamps.

Income tax.

The Income Tax Act (II of 1886) has not been applied to Baluchistān. British subjects in the service of the Government of India or of a local authority, or who may be serving within Native States in the Province, are alone liable to the tax. The receipts during the three years 1897 to 1900 averaged Rs. 15,400 per annum; in 1900–1 they were Rs. 16,200, and in 1902–3 Rs. 17,300.

Quetta is the only municipality that has been formally constituted in Baluchistān. An octroi tax was levied by Kalāt officials before British occupation, which was continued after Sir R. Sandeman's arrival, a conservancy cess being added in 1878. The site of the town and civil lines was purchased by Government, and was subsequently assigned to the municipality under certain conditions. Up to the year 1893 the affairs of the town and its funds were managed by the Extra Assistant Commissioner at Quetta, controlled by the Political Agent. The Quetta Municipal Law came into force on October 15, 1896. The municipal committee consists of a chairman and not less than six nominated members. The Political Agent is *ex officio* chairman, and the term of office of the members is one year. In March, 1904, the committee included five European *ex officio* members and eight natives. During the four years ending in 1901 the municipal income averaged 1·7 lakhs. The principal item is octroi, the receipts from which are about a lakh annually. Half of the net receipts from this source are paid over to the Quetta cantonment committee. Taking the population of the Quetta town alone, the incidence of taxation is Rs. 12–4–1 per head; including that of the cantonment, however, it drops to Rs. 6–11–10. Quetta being surrounded by open country, the cost of collection of octroi is necessarily costly, amounting to more than 19 per cent. of the collections in 1901. In 1903–4 the total income of the municipality had increased to 2·2 lakhs.

Local funds.

Seven 'excluded' Local funds have been created, known as the Sibi municipal fund, the Shāhrig District bazar fund (which includes the Ziārat improvement fund), the Pishīn Sadar and District bazar fund, the Loralai town fund (including the Duki and Bārkhān funds), the Fort Sandeman, the Nushki, and the Bolān bazar funds. These have been formed from time to time as new centres of trade sprang up and developed. The objects to which their income is devoted include local works and other measures of public utility, such as education and conservancy. The accounts, except in the case of the Sibi municipal fund, are governed by rules issued by the Government of India, and are audited by the Comptroller, India Treasuries. The Revenue Commissioner exercises with regard to their disposition the powers of a Local Government; the Political Agents are controlling officers, while the officers in charge of subdivisions (except in the Bolān, Pishīn, and Chaman) are the administrators. The total income of these funds during the four years ending in 1901 averaged Rs. 1,18,000 per annum, and the expenditure Rs. 1,14,000. No local rates are levied; but the principal source of income is octroi, which contributed an average of Rs. 62,700. One-third of the octroi receipts at Chaman, Loralai, and Fort Sandeman is paid over to the military authorities at these stations. The income of the funds in 1903–4 amounted to Rs. 1,07,000, and the expenditure to Rs. 1,01,000.

Quetta was declared a cantonment in 1883 and Loralai in 1897. The funds of each are administered by a cantonment committee. Their income consists chiefly of octroi and of grants-in-aid from Government, and is expended on objects similar to those of Local funds. During the four years ending in 1901, the total income averaged Rs. 93,000 and the expenditure Rs. 92,000 per annum. In 1903 the income amounted to Rs. 1,25,600 and the expenditure to Rs. 1,39,700. *Cantonments.*

The Public Works department has had a chequered career. Up to 1882 a Superintending Engineer was appointed under the immediate orders of the Local Government. From 1882 to 1885 the execution of civil works was entrusted to the Director-General of Military Works, the military Superintending Engineer being Secretary to the Local Government. From 1885 to 1889 public works were carried out by both military and civil agency, the civil Superintending Engineer being Joint Secretary to the Agent to the Governor-General. From 1889 till 1893 the civil Superintending Engineer was made Secretary for Public Works to the Local Government, and three civil divisions were created; the military Superintending Engineer was Joint Secretary and also controlled two Military Works divisions. *Public works.*

In 1893 all civil works were again entrusted to the military; and, owing to the importance of training Royal Engineer officers on the frontier and to the evils of a dual system, this arrangement still continues. Thus, the executive officers of *Present organization.*

the department of Military Works carry out both the civil and military works falling within their charges. The Commanding Royal Engineer, Quetta, is Superintending Engineer for all works and Secretary for Public Works to the Agent to the Governor-General. The Agency is divided into two sub-districts, each in charge of a Sub-Commanding Royal Engineer. Subordinate to the Sub-District Commanding Royal Engineers are Garrison Engineers, who have charge of local areas. All works carried out on behalf of the civil authorities are executed as contribution works to the Military Works department. For establishments, tools, and plant, and cost of audit a fixed sum of 24½ per cent. on the cost of all works is credited to the military works services. A payment of Rs. 2,300 is also made on account of establishments engaged for the super-vision of coal mines. A special Irrigation Engineer has been recently appointed, whose pay is debitable to Provincial revenues.

Powers of Engineer officers. The Commanding Royal Engineer, in his capacity as Superintending Engineer and Secretary to the Agent to the Governor-General, frames budget estimates, considers original major works costing more than Rs. 2,500, and allots sums for minor works. Sub-District Commanding Royal Engineers sanction original works costing not more than Rs. 2,500, which have been previously selected by the Governor-General's Agent, and dispose of the annual grants for repairs through Garrison Engineers. Minor works costing not more than Rs. 200 in each case are disposed of by civil officers.

Operations of the department. Military schemes. In the quarter of a century which has elapsed since its occupation, all the north-eastern part of the Province has been covered with a system of roads; bungalows and rest-houses have been built at convenient places, and water-supplies have been provided in the principal head-quarters stations. Many of the most important of these schemes owe their inception and execution to military needs. Such are the Bolān road; the buildings in the cantonments at Quetta and Loralai; and the water-supplies at Quetta, Sibi, and Loralai. The Pishīn-Dera Ghāzi Khān and Harnai-Loralai roads, though carried out as civil works, are now maintained from military funds.

Civil works. Of civil works, three canal systems, Shebo, Khushdil Khān, and the Anambār scheme, have been constructed at a total cost of 17.3 lakhs. An account of the most important roads and their cost has been given in the section on means of communication. Among public buildings may be mentioned the Quetta Hospital, which consists of fifteen blocks for in-patients and has a splendidly equipped operating room, the whole erected at a cost of about Rs. 76,000; the administrative buildings at Quetta in which the District courts and treasury are located, which were completed in 1892 at a cost of about 2 lakhs of rupees, including later additions; the Church of England, which cost 2·8 lakhs; and the Roman Catholic Church, to the erection of which a donation of Rs. 95,000 was given by Government. Two Residencies have been constructed for the Agent to the Governor-General. That at Quetta, which is one of the prettiest official residences in India, and cost 1·3 lakhs, was completed in 1893; the other is at Ziārat, the capital cost of which was about half a lakh. The Darbār Hall at Quetta, which contains a fine room for *darbār* purposes, was formerly used as the church, and cost about Rs. 92,000. The Public Works department has also constructed the Sandeman Memorial Hall at Quetta and the Victoria Memorial Hall at Sibi, in which the *Shāhi jirgas* are held. The former cost 1·1 lakhs, of which rather less than half was raised by private subscription, and the latter Rs. 38,800.

The only municipal drainage scheme of importance is that in Quetta town, on which about 1·1 lakhs have been expended on capital account. Permanent open drains have also been constructed in Sibi and Fort Sandeman. Piped water-supplies exist at Quetta, Fort Sandeman, Loralai, Sibi, Chaman, and Ziārat. *Municipal and cantonment schemes.*

Quetta was originally occupied in 1839, but was evacuated at the conclusion of the first Afghān War. On the outbreak of the second Afghān War in 1878 it was used as the base of operations, and troops held Pishīn, Quetta, and the line of the Bolān from Sibi. Loralai was occupied in 1886, and Chaman in 1889. Fort Sandeman was garrisoned in 1890. *Army.*

Quetta is the head-quarters of the fourth division of the Western Command. The troops in Baluchistān are under the divisional head-quarters direct, and a brigade under a colonel on the staff is located in Sind. The division is commanded by a major-general. The troops in the Province consisted in 1903 of three mountain batteries, two companies of garrison artillery, two British infantry regiments, three native cavalry regiments, six native infantry regiments, and one company of sappers and miners. The total strength of troops on June 1, 1903, was 2,650 British and 7,121 native; total 9,771.

The greater part of the garrison is quartered at Quetta and in a number of outposts; the remainder is distributed at Loralai, Fort Sandeman, and Chaman, each of these stations being garrisoned by native infantry and cavalry. In addition to the regular troops, a local company of volunteers and a company of North-Western Railway volunteers have been raised. They possessed 175 members on their rolls in 1891, 196 in 1901, and 236 in 1903.

Quetta itself is very strongly fortified by works which, with the support of the two lines of railway, render it practically impregnable. The fortifications were first designed in 1883, and have since been extended and improved. It contains an arsenal, and the Indian Staff College is in course of erection (1906).

The division possesses five local regiments, three of cavalry and two of infantry. The cavalry consists of the Scinde Horse, Jacob's Horse, and the Baloch Horse, which are borne on the Army List as the 35th, 36th, and 37th cavalry, and were raised in 1839, 1846, and 1885, respectively. The two infantry regiments are the 124th (Duchess of Connaught's Own) and the 126th Baluchistān Regiment.

Zhob Levy Corps. The Province has the distinction of possessing one of the first of those corps of local militia which now bear so large a part in frontier management. The Zhob Levy Corps was raised in 1890, and consists of four squadrons of cavalry, aggregating 423 men, and six companies of infantry, aggregating 632 men, under a commandant, second-in-command, and adjutant of the regular army[1]. It guards a long line of frontier from Loiband on the west to Gul Kach on the east, a distance of about 180 miles, besides garrisoning several posts to protect Zhob from the incursions of Mahsūd and other raiders. In 1902–3 the expenditure on the corps amounted to more than $2\frac{1}{2}$ lakhs. Details of the irregular forces maintained in KALĀT and LAS BELA will be found in the articles on those States.

The policy of making the inhabitants responsible for the security of the country and frontier has been developed into what is known as the levy system. Both levies and police are worked side by side; but the duties of the latter are, so far as possible, confined to guards and escorts, and to maintaining

---

[1] Since 1905 the corps has been strengthened by the addition of 200 mounted men and one British officer, its total strength being raised to 1,275 men.

order in the towns and on the railway. Crime is investigated
by the local headmen and levies, who are assisted by experienced deputy-inspectors of police and levies. This system
is supplemented by the reference of cases to *jirgas* for decision
by tribal custom. In 1904, 2,022 levies and 1,173 police were
employed in directly Administered areas, being one man to 109
persons or 15 square miles.

The levy system owes its origin to Sir Robert Sandeman, The levy
who, when Deputy-Commissioner of the Dera Ghāzi Khān system;
District in 1867, took into government service a small number its history.
of tribal horsemen from the Marris and Bugtis. These men
were employed chiefly in keeping up communications between
the chiefs and the British authorities. The system was extended, on Sir Robert Sandeman's arrival in Baluchistān, by
offering the headmen allowances for maintaining a certain
number of armed horse and foot, by whose means they were
expected to keep order in their tribes and to produce offenders
when crimes occurred. It was based on the assumptions,
firstly, that in every tribe there exist headmen of influence
who, if effectually supported, can compel obedience, and,
secondly, that no frontier tribesman can be expected to work
for Government unless paid for it. The system thus
started was expanded in 1883, when the organization was
fully considered by a committee. At this time the Baloch
Guide Corps was disbanded. It had been raised for service
in Kachhi at the time of the first Afghān War in 1838, but
was withdrawn within the Sind frontier in 1852 and transferred to Baluchistān in 1877[1].

The District officers control the levies, whose duties are Organiza-
many and various. Besides the detection and arrest of crimi- tion.
nals, they guard communications, bring in witnesses, make
inquiries, carry letters, produce supplies, escort prisoners, and
assist in the collection of revenue. They are exclusively
recruited from the local tribes, and are ordinarily employed
within their own tribal limits. Each man has his own weapon,
generally a sword, and is distinguished by a badge or turban.
In some places the men are armed with Snider carbines. At
central posts writers are stationed to conduct correspondence.
The system cannot be gauged by the actual numbers employed,
as the grant of service entitles the Government to the assistance of the whole tribe in times of emergency and not merely

---

[1] For a short account of the duties performed by this corps in Sind see
Baluchistān Blue Book No. 2, p. 70.

to that of the few enrolled individuals. It is also an axiom that all persons drawing pay, whether chiefs or others, who are not pensioners, must render an equivalent in service. A few personal grants have been made for long and tried service, but these are exceptional. Reference has already been made to the condition in the Provincial settlement that the expenditure on levies shall not be materially reduced below Rs. 8,27,000. This sum includes the cost of the Zhob Levy Corps, the expenditure on which amounts to more than $2\frac{1}{2}$ lakhs of rupees. Full statistics of the force of levies employed at different periods will be found in Table II at the end of this article. In 1904 the total number of effective levies working in the Province, excluding the Zhob Levy Corps and those employed on postal duty, amounted to 2,130, and cost about Rs. 5,86,000 per annum.

Police.
Police have been recruited from time to time as necessity demanded. In 1882 the total number of police was 516, but no special officer had been appointed to supervise them. In 1889 the question of police administration was considered by a committee ; but the system then initiated, by which the control of the force was handed over to Political Agents and Assistant Political Agents, was altered in 1897, when a European District Superintendent of police was appointed to Quetta-Pishīn and the railway. He has since been placed in charge of the Sibi District police. The Agent to the Governor-General is the Inspector-General of Police and Levies. In Zhob and Loralai an Inspector of police, who is also an Honorary Assistant District Superintendent, holds charge of the force. Statistics of the total force at different periods will be found at the end of this article. In 1904 the total strength of the police was 1,173 of all grades, or one policeman to 41 square miles of directly Administered areas. The number of officers, inspectors, and deputy-inspectors was forty-two.

The police are chiefly recruited from natives of India, but local men are now beginning to join. A tendency exists among the latter not to serve continuously for long periods. Efforts are being made to enlist local men of good family in the higher grades. Anthropometrical measurements and finger-prints are not taken. The armament has hitherto consisted of Snider rifles with side-arms, for which Martini rifles are to be substituted. Much use is made of trackers, who frequently exhibit remarkable talent. The railways are divided into two subdivisions for police purposes. The number of police employed on them is 147, and they are posted at 42 stations. This number is supplemented by 160 levies, making an aggregate of 307 men employed on 399 miles of line.

During the three years ending 1901 the average number of cognizable cases reported was 1,542. Of these cases 1,055 were decided in the criminal courts, 948 ending in conviction and 107 in acquittal or discharge. Cases of serious crime, which would ordinarily be treated as cognizable in India, are, for political reasons, not always investigated by the police in Baluchistān, but are placed before tribal *jirgas* for award. Cognizable crime.

The Revenue Commissioner is Inspector-General of Jails. The Province possessed two District jails and fourteen subsidiary jails in 1900–1, capable of accommodating 490 male and 93 female prisoners. Details of the prison population, &c., are given in the following table :— Jails.

Year.	District jails.	Subsidiary jails (Lock-ups).	Average daily jail population.			Rate of jail mortality per 1,000.	Expenditure on jails and maintenance of prisoners.	Cost per prisoner.
			Males.	Females.	Total.			
							Rs.	Rs.
1891–2	1	5	*93	†3	*96	41	12,831	133
1895–6	2	12	*127	†2	*129	†42	14,320	104
1900–1	2	14	247	11	258	†47	23,034	96

* Includes figures for the Sibi and Harnai jails only in the Thal-Chotiāli (now Sibi) District.

† Includes figures for the Sibi subdivision only in the Thal-Chotiāli (now Sibi) District.

NOTE.—No figures are available for the Bolān Pass.

The principal causes of sickness among the prisoners are malaria, dysentery, and pneumonia in winter. The diet is much the same as in the Punjab, the Jail Manual of which Province is followed *mutatis mutandis*. Prisoners whose term exceeds six months are ordinarily sent to Shikārpur in Sind, under arrangement with the Bombay Government.

No industries are carried on in the local jails, except in Quetta and Sibi, where coarse blankets are woven for the bedding and clothing of prisoners. Juvenile prisoners are sent to the reformatory at Shikārpur.

Education has as yet made little progress, and a department of Public Instruction is only now being organized by an Education.

Inspector-General of Education appointed jointly for the Frontier Province and Baluchistān. Secondary education is represented by one high and one Anglo-vernacular middle school. The number of boys in these two schools was 27 (1891), 103 (1901), and 87 (1903). They are principally maintained from Local funds, but receive pecuniary aid from Government.

Primary education.
The primary schools attached to the middle schools are Anglo-vernacular, in so far that English is taught in the two higher classes of the primary department. In all other primary schools the medium of instruction is Urdū, and the subjects are Urdū, Persian, geography, and arithmetic. The number of boys' schools has been three (1891), fourteen (1901), and twenty-one (1903). They contained 604 pupils in 1901 and 831 in 1903. Difficulties are experienced in obtaining qualified teachers, who generally have to be recruited from the Punjab. Their pay varies from Rs. 25 to Rs. 35 per mensem.' A rough estimate shows that the number of pupils under instruction in mosque schools in 1901 was 2,256 boys and 313 girls.

Female education.
The number of female schools and pupils has been one school with 61 pupils (1891); three schools with 170 pupils (1901); and four schools with 240 pupils (1903). Of the four schools, two are maintained from Local funds and two by private bodies. Of the latter, one is aided from Local funds and the other is a mission school. Each school is divided into three departments, Hindī, Gurmukhī, and Urdū, these languages being the medium of instruction. The Punjab system is followed, and sewing and knitting receive special attention. Almost all the girls attending these schools are from India. In 1901 the number of Hindu girls represented 82 per cent., and in 1903 72 per cent. of the total. The usual difficulties caused by early marriage and the *parda* system are the great hindrances to progress.

Europeans, Eurasians, and Muhammadans.
Two European schools contained thirty-one pupils (1901) and fifty-three (1903). Teaching on the Punjab system is given up to the middle department in one; the other is a primary school. In 1901 the total number of Muhammadan children under instruction was 434, including 28 girls. Indigenous children numbered 227. In 1903 the number of Muhammadan pupils had risen to 548, including 349 indigenous children.

General results.
The native population is too poor and uncivilized to appreciate the benefits of education, but special inducements

are held out to them in the shape of stipends. The number of boarding-houses has risen from one accommodating nineteen residents (1901) to two accommodating thirty-two residents (1903). The census statistics afford no indication of the progress made in education. The Province is still very backward, absence of funds being the greatest drawback. It has not yet been found possible to charge fees in village primary schools. Educational expenditure during 1903-4 is exhibited in the table below:—

	Provincial revenues.	Local and municipal funds.	Fees and other sources.	Total.
	Rs.	Rs.	Rs.	Rs.
Secondary boys' schools	4,890	3,974	3,840	12,704
Primary ,, ,,	4,585	4,334	3,198	12,117
Girls' schools	180	4,534	2,468	7,182
Total	9,655	12,842	9,506	32,003

The *Baluchistān Gazette* is the only newspaper published in English, and is chiefly devoted to local news. No vernacular newspapers exist, and no books have been issued. Native Press.

Dispensaries were first opened at Kalāt and Quetta and a Medical Officer was appointed to the Agency in 1877. The department has since rapidly expanded. Statistics of the number of civil hospitals and dispensaries and of the patients treated are given below:— Dispensaries and diseases.

YEAR.	Number of hospitals and dispensaries.	AVERAGE DAILY NUMBER OF PATIENTS.	
		Indoor.	Outdoor.
1891	12	86	525
1896	15	113	659
1901	16	113	703
1903	17	120	725

These figures exclude the statistics for institutions maintained by the North-Western Railway, private bodies, and Native States, which numbered seven (1891), nine (1896), ten (1901), and eleven (1903). The total number of cases treated in all institutions has been 8,590 (1881), 112,809 (1891), 207,534 (1901), and 204,611 (1903). In 1887 the system of medical relief was remodelled, all medical institutions being placed under the Agency Surgeon and Administrative Medical Officer, who was made the central controlling authority. A Civil Surgeon is stationed at Quetta, and the military European medical officers at garrison stations have been appointed Civil Surgeons of those places. Seven Assistant Surgeons and twenty-eight Hospital Assistants are employed. The largest hospital is at Quetta. Two hospitals are supported by the Dufferin Fund.

The principal source of income is a grant from the Provincial revenues. The cost of medicines, diet, &c., has largely increased in recent years. The following figures indicate the income and expenditure of civil medical institutions maintained by Government :—

YEAR.	INCOME.				EXPENDITURE.		
	Total.	Government payments.	Local and municipal payments.	Fees, endowments, and other sources.	Total.	Establishments.	Medicines, diet, buildings, &c.
	Rs.	Rs.	Rs.	Rs.	Rs.	Rs.	Rs.
1891 . .	27,700	22,500	3,600	1,600	26,000	18,800	7,200
1896 . .	29,600	25,400	3,800	400	30,800	21,200	9,600
1901 . .	43,100	35,700	6,700	700	42,900	22,900	20,000
1903 . .	59,300	52,800	5,100	1,400	62,000	24,400	37,000

The commonest endemic diseases are malarial fevers and cachexia, bowel complaints, scurvy, chronic ulcers including 'frontier sores,' chronic inflammation of the eyes, dysentery, and rheumatism. They are caused chiefly by insufficient clothing, insanitary dwellings, poor diet, the absence of anti-scorbutics, and the dust and dryness of the atmosphere. Of epidemics small-pox, measles, cholera, and typhus fever are the most prevalent.

Ten outbreaks of cholera have occurred between 1887 and 1903. Contrary to general rule, it is quite possible to secure effective land quarantine in Baluchistān. There have been four known outbreaks of typhus fever since 1891, but the occurrence of this disease appears to have remained unidentified in several instances. An inspection camp established on the railway in 1896 has hitherto (1905) prevented the spread of plague to Upper Baluchistān, but an outbreak occurred in 1902 at Sonmiāni in Las Bela.

Lunatics. The Province possesses no lunatic asylum. Lunatics are sent to the Hyderābād Asylum in Sind. The following table shows the number of criminal and other lunatics from Baluchistān and the cost of each inmate :—

	1891.	1896.	1901.	1903.
1. Average daily number of —				
(*a*) Criminal lunatics .	None	None	0.6	1.5
(*b*) Other lunatics .	7.3	13.9	6.1	12.3
Average cost of each inmate per annum . .	Rs. a. p. 70-7-9	Rs. a. p. 103-9-10	Rs. a. p. 109-6-5	Rs. a. p. 104-8-6

Lunatics from Baluchistān are reported to be generally well-behaved, but to be specially subject to outbursts of passion and acts of violence.

Since the introduction of vaccination the outbreaks of Vaccination. small-pox have been limited in both extent and virulence. Vaccination is available throughout directly Administered areas, but is compulsory only in Quetta. Inoculation is freely practised by *mullās*, Saiyids and others, a small fee, generally four annas, being charged for the operation. The people are still ignorant and apathetic, and rarely resort to vaccination till an outbreak of small-pox occurs. Re-vaccination is seldom permitted. In 1896, 12,500 operations were performed, of which 11,000 were successful ; in 1903 the number was 13,000, of which 10,500 were successful. The cost per successful case was two annas and two pies in 1896, and four annas and eight pies in 1903.

The sale of quinine in pice packets is progressing : 6,627 Village packets were sold in 1895–6 ; 8,694 in 1901–2 ; and 8,050 in sanitation. 1902–3. Village sanitation is conspicuous by its absence. In the highlands manure is freely used in cultivation, and sweepings and dirt are thus removed to some extent from the neighbourhood of habitations. The nomads move their encampments when their surroundings become filthy beyond endurance.

The coast of the Province was originally surveyed by officers Surveys. of the late Indian Navy between 1823 and 1829. These surveys were afterwards revised and corrected by Lieutenant A. W. Stiffe in 1874, and excellent charts have been published by the Admiralty. The charts are supplemented by a very complete and accurate account of the coast known as the *Persian Gulf Pilot*[1].

[1] London, 1898. Sold by J. D. Potter, 31, Poultry, and 11, King Street, Tower Hill.

A systematic survey of the interior was commenced by the Survey department of the Government of India in 1879, and during the succeeding twenty years a triangulation connected with the Indus series of the Great Trigonometrical Survey was extended through the country. In spite of great difficulties, topographical surveys on the half-inch scale have been carried out up to the 66th east parallel, and maps have been published on that scale. Westward, the results of a reconnaissance survey have been embodied in published maps on the quarter-inch scale. Several special large-scale surveys have also been undertaken, including one for a railway in the Zhob valley and another of the coal-bearing Sor range.

Settlement surveys. In connexion with the settlement operations a cadastral survey of all irrigated villages in the Quetta, Pishīn, Shāhrig, Sanjāwi, Sibi, Hindubāgh, Kila Saifulla, and Bori *tahsīls* has been undertaken on the scale of 16 inches to 1 mile. These surveys are based on traverses carried out by the Survey of India department, and are being extended to the Duki and Bārkhān *tahsīls*. The agency employed is almost entirely foreign, and the local Afghān has so far shown little aptitude for acquiring the principles of surveying.

Bibliography. [*Baluchistān Blue Books*, Nos. 1 and 2 (1877.)—*Baluchistān Agency Annual Administration Reports*. (Calcutta.) — *Census of India*, 1901, vols. v, v a, v b.—G. B. Tate: *Kalāt*. (Calcutta, 1896.)—'Geographical Sketch of the Baluchistān Desert and part of Eastern Persia': *Memoirs of the Geological Survey of India*, vol. xxxi, pt. 2. (Calcutta, 1904.)—M. L. Dames: *A Historical and Ethnological Sketch of the Baloch Race* (1904).—Sir T. H. Holdich: *The Indian Borderland* (1901).—A. W. Hughes: *The Country of Baluchistān* (1877).—J. H. Lace: 'A Sketch of the Vegetation of British Baluchistān': *Journal Linnean Society*, vol. xxviii (1891).—C. Masson: *Narrative of a Journey to Kalāt* (1843); *Journeys in Balochistān, Afghānistān, and the Punjab* (1842).—A. H. McMahon and Sir T. H. Holdich: Papers on the Northern and Western Borderlands of Baluchistān in *Geographical Journal*, vol. ix.—H. Pottinger: *Travels in Beloochistan and Sinde* (1816).—T. H. Thornton: *Life of Colonel Sir Robert Sandeman* (1895).]

TABLE I.

ADMINISTRATIVE DIVISIONS.	Area in square miles.	Number of villages.	TOTAL POPULATION, 1901.			Persons per square mile in rural areas.
			Persons.	Males.	Females.	
I. BRITISH AND ADMINISTERED TERRITORY.						
(1) Zhob . . .	9,626	245	69,718	39,637	30,081	7
(2) Loralai . . .	7,999	400	67,864	37,823	30,041	8
(3) Quetta-Pishīn . .	5,127	329	114,087	68,945	45,142	17
(4) Chāgai, including Western Sinjrāni * .	18,892	32	21,689	11,418	10,271	1
(5) Bolān . . .	896	8	1,936	1,483	453	2
(6) Sibi. . . .	4,152	304	73,893	41,619	32,274	17
Total	46,692	1,318	349,187	200,925	148,262	7
II. NATIVE STATES. A. Kalāt State.						
(1) Sarawān . . .	4,339	298	65,549	36,366	29,183	14
(2) Kachhi . . .	5,310	606	82,909	44,836	38,073	16
(3) Jhalawān . .	21,128	299	224,073	115,077	108,996	11
(4) Khārān * . . .	14,210	20	19,610	8,825	10,785	1
(5) Makrān * . . .	26,606	125	78,195	35,188	43,007	3
Total	71,593	1,348	470,336	240,292	230,044	7
B. Las Bela State .	6,441	139	56,109	29,718	26,391	9
Total	78,034	1,487	526,445	270,010	256,435	7
III. TRIBAL AREAS.						
(1) Marri Country . .	3,268	5	20,391	11,491	8,900	6
(2) Bugti Country . .	3,861	3	18,528	10,266	8,262	5
Total	7,129	8	38,919	21,757	17,162	5
GRAND TOTAL	131,855	2,813	914,551	492,692	421,859	5

* The figures for Makrān, Khārān, and Western Sinjrāni are estimates made in 1903.

Note.—There were six towns in Baluchistān in 1901; one each in the Sibi, Loralai, and Zhob Districts, and three in Quetta-Pishīn. The urban population numbered 40,033, being 31,757 males and 8,276 females.

TABLE II

Levies

Year	Total number and cost of levies		Number of horsemen.	Number of footmen.	Native officers (Daffadárs and upwards).	Number of headmen and others receiving allowances.	Number of men receiving political and charitable allowances.	Miscellaneous, clerks, &c.
	Number.	Cost.						
		Rs.						
1891–2	1,994	4,02,725	1,002	502	293	55	74	68
1895–6	1,989	5,10,363	985	480	303	67	65	89
1900–1	2,337	6,06,088	1,144	537	366	90	90	110
1902–3	2,351	6,02,747	1,164	532	354	95	94	112

* Including postal and telegraph levies, but excluding Zhob Levy Corps.

Police

Year.	Total.	DISTRICT POLICE.							Municipal and cantonment police.		Railway police.		Total cost.	Watchmen paid from Local funds.	
		Supervising staff.			Subordinate staff.										
		District and Assistant Superintendents.	Inspectors.	Deputy Inspectors.	Sergeants.	Mounted officers.	Mounted constables.	Footmen.	Supervising staff.	Subordinate staff.	Supervising staff.	Subordinate staff.		Number.	Cost.
													Rs.		Rs.
1891–2	1,005	...	2	18	131	13	114	445	...	118	2	162	1,84,189	41	5,123
1895–6	1,082	1	2	20	152	12	122	508	...	127	2	136	2,37,972	101	12,672
1900–1	1,159	3	3	21	161	12	125	557	1	138	2	136	2,80,179	98	11,998
1902–3	1,186	4	3	20	162	12	124	562	1	167	1	130	2,71,683	94	11,221

Brāhuis, The.—A confederacy of tribes occupying the
Sarawān and Jhalawān country of the Kalāt State in Baluchistān, and headed by the Khān of Kalāt. The Brāhuis are
divided into two main divisions, each under its own leader :
the Sarawāns living to the north of Kalāt under the Raisāni
chief, and the Jhalawāns to the south under the Zahri chief.
The Sarawān division includes among its principal tribes the
Raisāni, Shāhwāni, Muhammad Shāhi, Bangulzai, Kūrd, Lehri,
and Sarparra. The Lāngav, though not occupying a position
of equality with those just named, are also reckoned among
the Sarawāns. Among the Jhalawāns are the Zahri, Mengal,
Mīrwāri, Bīzanjau, Muhammad Hasni or Māmasani, and several
others. At the head of each tribe is a chief, who has below
him subordinate leaders of clans, sections, &c. The whole
tribe is united by common blood-feud rather than by kinship.
When occasion arises intersectional combinations take place.
The internal administration of each tribe is independent, cases
being settled by the chief in consultation with his headmen.
The crystallization of the tribal groups into the Brāhui confederacy was completed by Nasīr Khān I, each tribe being
bound to furnish a number of armed men, and intertribal
cases being referred to the ruler. That the Brāhuis are
essentially nomads and flockowners is well indicated by their
proverb : ' God is God, but a sheep is a different thing.' The
Muhammadan religion which they profess is largely overgrown
with animistic superstitions. Hospitality is common, but is not
so profuse as among the Baloch.

The origin of the Brāhuis is as much an enigma to the ethno-
logist as their language has been to the philologist. The theory
that their name is derived from the old Persian words *ba
rohi*, ' a hillman,' may be rejected. Their own, and the most
plausible, explanation is that the word Brāhui is derived from
the eponym of one of their forefathers, *Brāho*, which is a not
uncommon modification of the name *Brāhīm* or *Ibrāhīm* at
the present day. Early Baloch poems also describe them as
the *Brāho*. In the light of anthropometrical measurements
recently made, Mr. Risley classes the Brāhuis as Turko-Irānians.
It seems not unlikely that they also contain remnants of those
hordes of broad-headed nomadic people who came into India
at the beginning of the Christian era and are known by the
generic term of Scythians. We first find the Brāhuis in
authentic history divided into groups clustering round Kalāt
under a chief drawn from their senior branch, the Mīrwāris,
and called Mīr Umar. Driving out the Jat population of the

Jhalawān country, they made themselves masters of the whole country between Mastung and Las Bela. Only Mīr Umar's descendants are now regarded as true Brāhuis. They include the Ahmadzais, the ruling family, with their collaterals the Iltazais; the Sumalānis, Kalandarānis, Gūrgnāris, Kambarānis, Mīrwāris, and Rodenis. As the power of the chiefs expanded, the name Brāhui was extended to the various groups which were included in the confederacy from time to time, numbers of Jats, Afghāns, and Baloch being thus absorbed.

The Brāhui is of middle size, square-built and sinewy, with a sharp face, high cheek-bones, and long narrow eyes. His nose is thin and pointed. His manner is frank and open. Though active, hardy, and roving, he is not comparable with the Baloch as a warrior, but he makes a good scout. The songs and ballads of the people celebrate no days on which hundreds were killed, as in the case of the latter. With few exceptions the Brāhui is mean, parsimonious, and avaricious, and he is exceedingly idle. He is predatory but not a pilferer, vindictive but not treacherous, and generally free from religious bigotry. His extreme ignorance is proverbial in the country-side: 'If you have never seen ignorant hobgoblins and mountain-imps, come and look at the Brāhui.' The Brāhui wears a short smock descending to the knees and fastening on the right shoulder, wide trousers often dyed black or brown, and a felt cap or a turban. His foot-covering consists of sandals or embroidered heavy shoes. He is fond of having a waistcoat over his smock, and he also wears a black overcoat (*shāl*). A woman's dress consists of a long shift profusely embroidered in front. If married, she wears a kind of corset lacing behind. Her hair is done in two plaits joined at the back and covered by a long cotton scarf.

The Brāhui language has long been an interesting puzzle to the philologist. Like the Basque of Europe it stands alone among alien tongues, a mute witness to ethnical movements occurring before the rise of authentic history. It has no literature of its own, and our limited knowledge of it is due to European scholars. Some have connected Brāhui with the Aryan group, others with the Kol language of Central India; while others, among whom is Dr. Trumpp, place it with the Dravidian tongues of Southern India. Dr. Caldwell refused Brāhui a place in his list of Dravidian tongues, though he admitted that it contained a Dravidian element. The latest inquiries, however, confirm its connexion with Dravidian. Among its most striking points of likeness to the South-Indian group are some of its pronouns and numerals, the use of post-positions instead of prepositions, the absence of a comparison of adjectives by suffixes, the lack of the relative pronoun except as borrowed, and the negative conjugation of the verb.

Sulaimān Range (28° 31' to 32° 4' N. and 67° 52' to 70° 17' E.).—A range of mountains in North-Western India, about 250 miles long, lying between the Gomal river on the north and the Indus on the south, and separating the Frontier Province and the Punjab from Baluchistān. Its backbone consists of a main ridge running north and south, flanked on the east by parallel serrated ranges. On the Baluchistān side these flanking ranges gradually take an east and west direction to meet the Central Brāhui range. The height of the range gradually decreases to the southward. The geological formation of the southern parts is distinct from that of the northern. In the former, sandstones, clays, and marls predominate; in the latter, pale marine coral limestone rests on cretaceous sandstone. Petroleum has been worked in the Marri hills. On the southern slopes vegetation is scarce; in the central part olives abound; farther to the north the higher elevations are covered with edible pine (*chilghoza*), the fruit of which is collected and sold. In this part of the range much magnificent scenery is to be found, of which the extraordinarily narrow gorges constitute the most striking feature. These clefts afford a means of communication with the Punjab, the principal routes being through the Ghat, Zao, Chuharkhel Dhāna, and Sakhi Sarwar Passes. The highest point of the range, 11,295 feet above the sea, is known to Europeans as the Takht-i-Sulaimān ('Solomon's throne') and to natives as Kasi Ghar. Sir Thomas Holdich[1] describes the *takht* as a *ziārat* or shrine situated on a ledge some distance below the crest of the southernmost bluff of the mountain. It is difficult of approach, but is nevertheless annually visited by many pilgrims, both Hindu and Muhammadan. The inhabitants in the northern parts of the range are Afghāns, and in the south Baloch. About thirty miles north-west of Fort Sandeman lies the picturesque little sanitarium of Shīnghar. Farther south is the Punjab hill-station of Fort Munro (6,363 feet) in Dera Ghāzi Khān District. Straight-horned *mārkhor* (*Capra fal-coneri*) are to be found at the higher and mountain sheep (*Ovis vignei*) at the lower elevations.

Toba-Kākar Range.—A mountain range (30° 22' to 32° 4' N. and 66° 23' to 69° 52' E.) in the Zhob and Quetta-Pishīn

---

[1] *The Indian Borderland*, p. 73.

Districts of Baluchistān, which forms the boundary between
Baluchistān and Afghānistān, and at the same time the watershed
between India and Central Asia. It is an offshoot of the Safed
Koh, with three parallel ridges gradually ascending in a south-
westerly direction from a height of about 5,000 feet near the
Gomal to the peaks of Sakīr (10,125 feet), Kand (10,788 feet), and
Nigānd (9,438 feet) in the centre. Thence it descends towards
the west and, opposite Chaman, takes a sharp turn to the south-
west, continuing under the name of the Khwāja Amrān and Sarlath. Eventually it merges into the Central Makrān range, after
a total length of about 300 miles. The country between the
Gomal and the Kand peak, which is drained by the Kundar and
Zhob rivers, is known from its inhabitants as Kākar Khorāsān.
The part to the westward of the Kand peak is called Toba, and
is inhabited chiefly by Achakzai Afghāns. The range has never
been entirely surveyed. The higher elevations consist of wide
plateaux, intersected on either side by deep river valleys. In
winter the cold on these wind-swept plains is intense. They
are covered thickly with the small bushy plant called southernwood (*Artemisia*). Little timber is to be seen. Bosomed in
the Kand mountain is one of the most picturesque glens in
Baluchistān, called Kamchughai. Across the Khwāja Amrān
offshoot lies the KHOJAK Pass. Another important pass in the
Khwāja Amrān is the Ghwazha. The most interesting feature
of the geology of the range is the continuation of the Great
Boundary Fault of the Himālayas which runs along it. The
upper strata consist of flysch, known to geologists as Khojak
shales, beneath which lies a conglomerated mass of shaly bands
and massive limestone. Intrusions of serpentine, containing
chrome ore and asbestos, also occur.

Central Brāhui Range.—A mountain range in Baluchistān,
occupying the northern part of the Jhalawān and the whole of
the Sarawān country in the Kalāt State and part of the Administered areas, and forming the upper portion of the great system
to which Pottinger[1] gave the name of the Brahooick Mountains. It lies between 27° 57' and 30° 36' N. and 66° 31' and
67° 52' E., including the whole of the mass of mountainous
country between the Mūla on the south and the Pishīn Lora and
Zhob rivers on the north. Between the Mūla river and Quetta
the strike is north and south, but a few miles north of the latter
place the ranges turn sharply to the east and south-east to meet
the Sulaimān Mountains. The total length of the arc thus
formed is about 225 miles and the breadth about 70 miles.

[1] *Travels in Beloochistan and Sinde*, p. 251.

The general formation is a series of parallel ranges containing in their midst those narrow valleys which form the upper highlands of Baluchistān. All the highest peaks in the Province are situated in this system. They include Khalīfat (11,440 feet), a magnificent mountain having a sheer drop of 7,000 feet to the Shāhrig valley; Zarghūn to the north of Quetta (11,738 feet); Takatu (11,375 feet); the Koh-i-mārān (10,730 feet); and the Harboi hill, the highest point of which is Kakku (9,830 feet). None of the ranges has an altitude of less than 6,000 feet. They are composed chiefly of massive limestone—well exposed in Takatu and Khalīfat—which passes into an enormous thickness of shales. Zarghūn consists of conglomerate belonging to the Siwālik series. Coal is found towards Harnai. The southern parts of the range are inhabited by tribes of Brāhuis, while to the north live Afghāns, chiefly Kākars. Near the north-east end of the range lies ZIĀRAT, the Provincial summer headquarters. The railway traverses the Bolān and Harnai Passes. Another important pass is the Mūla. Unlike most of the other ranges of the Province, the Central Brāhui range is comparatively well clothed with vegetation, especially the Ziārat, Zarghūn, and Harboi hills. All the principal 'reserved' forests in Administered areas are situated on it. Juniper is most abundant, the trees being of great age; but the largest grow to a height of only about sixty or seventy feet. The timber is used for fuel and in a few places for building purposes.

Kīrthar Range.—A mountain range forming the boundary between Sind and the Jhalawān country in Baluchistān, between 26° 13′ and 28° 36′ N. and 67° 11′ and 67° 40′ E. From the point where the Mūla river debouches into the Kachhi plain, the range runs almost due south for a distance of 190 miles in a series of parallel ridges of bare rocky hills. At intervals similar ranges run athwart them. The offshoots tail off south-eastwards into Karāchi District, but a single line of low hills extends as far as Cape Monze. The greatest breadth is about sixty miles. The highest point is the Zardak peak (7,430 feet), and another fine peak is the Kuta-ka-kabar, or Kuta-jo-kabar, i.e. 'the dog's tomb' (6,878 feet). The principal offshoot is the Lakhi range. The Kīrthar hills are pierced by the Kolāchi or Gāj river in a fine gorge, and the chief passes are known as the Harbāb, Phusi, Rohel, and Garre. These hills give their name to the Kīrthar geological group of nummulitic limestone, which is found on their crests, overlaid by tertiary rocks of Nāri and Gāj beds, the former being soft sandstone and the latter a hard dark-brown limestone exposed on the

Gāj river. The tribes residing in the Kīrthar are the Marri and Jamāli Baloch, Jamot and Chuta Jats, and some Khidrāni and Sāssoli Brāhuis. They subsist chiefly by tending flocks, and by exporting the dwarf-palm (*Nannorhops Ritchieana*). Sind ibex and mountain sheep are fairly plentiful, and both black bears and leopards are occasionally met with.

Pab.—A range of mountains in Baluchistān, occupying the south-eastern corner of the Jhalawān country and the Las Bela State, between 24° 53′ and 27° 36′ N. and 65° 50′ and 67° 11′ E. Its general direction is north and south from the centre of the Jhalawān country to the sea ; on the east it is bounded by the valley of the Hab river, and on the west by the Hingol and its tributary the Arra. Locally, the name Pab is applied only to the backbone of the range ; but the generic term may appropriately be extended to the whole mass, which is in the form of a cow's udder, the Khude or Khudo, Mor, and Hālā hills, together with the backbone already mentioned, forming the teats. Within these ranges, reckoning from east to west, lie the valleys of Sārūna and Kanrāch and the Las Bela plain. The drainage is carried off by the Porāli, Hab, and Hingol rivers. The greatest length of the range is 190 miles and the breadth about 70 miles. The highest point, Pharās, is 7,759 feet above sea-level. The numerous limestone ridges generally rise precipitously on their eastern sides, but possess a more gradual slope to the west. The inhabitants are chiefly Mengal Brāhuis, with some Jāmots and Bhākras in the south. Besides flock-owning, their principal means of subsistence is the export of *pīsh* (*Nannorhops Ritchieana*) to Sind. Olive grows in the higher parts and acacia in the lower. Rich grazing grounds for camels, sheep, and goats occur. Copper is reported to be obtainable in the Mor hills. The principal passes are the Lār Lak, Chūri Pass, Bārān Lak, and Jau Lak. Towards the southern extremity of the range, in Las Bela, is the shrine of Shāh Bilāwal.

Chāgai and Rās Koh Hills.—The Chāgai hills (28° 46′ to 29° 34′ N. and 63° 18′ to 64° 50′ E.) are a range of mountains in the north of the District of the same name in Baluchistān. They have never yet been entirely explored. Their general strike is east and west, the main mass being about 90 miles long by 35 broad, lying to the west of the Hāmūn-i-Lora at a mean elevation of about 6,000 feet. The range extends farther westward, however, in a series of scattered volcanic ranges, the chief of which is the Koh-i-Sultān. The principal peaks of the main mass are Mārān (7,309 feet), Malik Teznān

(7,686 feet), and Malik Nāru (7,915 feet). Here and there are open plains containing slight cultivation and probably capable of development. The geological structure consists chiefly of basic and acid igneous rocks, with occasional outcrops of quartz and diorite. Terraces of travertine or Oriental alabaster occur at the western end of the main mass. The inhabitants are few and are principally nomad Baloch. The Koh-i-Sultān, which is separated from the main range by a sandy plain, is an oval-shaped mountain about 20 miles long by 14 wide. It is an extinct volcano with three distinct cones. Its most remarkable feature is the Neza-i-Sultān, 'the sultan's spear,' a huge natural pillar about 300 feet in diameter and 800 feet high. West of the Koh-i-Sultān rises another extinct volcano called Damodim. On the south, the Chāgai hills are connected with the Rās Koh hills by the Dālbandin plain. The latter range lies between 28° 25′ and 29° 13′ N. and 63° 57′ and 66° 0′ E. It is an extension, about 140 miles in length, of the Khwāja Amrān offshoot of the Toba-Kākar range, and takes its name from its highest peak, nearly 10,000 feet above sea-level. It gradually sinks westward beneath the superficial deposits of the Western Sinjrāni desert. The watershed roughly divides Khārān from Chāgai. Though the general direction is north-east to south-west, the component ridges have many irregularities in strike. Geologically also the formation is complex, consisting now of tertiary limestone, now of flysch, and now of igneous outcrops, which are best represented by the Rās Koh peak. Other peaks are Shaikh Husain (6,875 feet) where a shrine of some local importance exists, and Kambrān (8,518 feet). Vegetation is extremely scanty and the ridges rise bare and barren. Good bags of Sind ibex are sometimes made on them.

Siāhān.—A mountain range in Baluchistān, separating Makrān from Khārān. The eastern part is known as Band. It runs south-south-west and east-north-east between 27° 7′ and 28° 2′ N. and 63° 22′ and 65° 42′ E., and unites with the Jhalawān hills near Shireza, having a total length of 176 miles. It is the narrowest range in Western Baluchistān, the width nowhere exceeding 20 miles. North of Panjgūr the general mass bifurcates, the spur on the south being known as the Koh-i-Sabz. Its general aspect is abrupt and rugged, and its geological formation a slaty shale. It has a mean elevation of about 5,000 feet. On the west are the two fine defiles of Tank-i-Grawag and Tank-i-Zurrati, through which the Rakhshān river passes.

Central Makrān Range.—A mountain range in Baluchistān, occupying the centre of Makrān, between 26° 3′ and 27°
39′ N. and 62° 19′ and 65° 43′ E. Springing from the hills
of the Jhalawān country its two well-defined and gradually
descending ridges, the Zangi Lak or Dranjuk hills (6,166 feet)
on the north and the Koh-i-Patandar (7,490 feet) with its continuation the Kech Band (3,816 feet) on the south, run west-
south-west for about 250 miles. The tumbled mass in the
centre merges on the west into the Zāmurān hills, and the
northern portion stretches into the Persian Bāmpusht range.
The width is uniform, about 45 miles. Sandstone is the
prevailing rock, sometimes associated with shaly strata and
limestone. Within the range lie the valleys of the Rāghai,
Gichk, and Gwārgo rivers, Bālgattar, Buleda, and Parom. The
Zāmurān hills are alone inhabited and have some cultivation
and vegetation.

Makrān Coast Range.—A mountain range in Baluchistān,
known locally as Bahr-i-Garr, which skirts the Arabian Sea for
280 miles between 25° 22′ and 26° 0′ N. and 61° 44′ and 66°
3′ E. Its width varies from 35 to 70 miles. The prevailing
rock is a pale-grey clay or marl, occasionally intersected by
veins of gypsum and interstratified bands of shelly limestone and
sandstone. The parallel ranges of the system descend gradually
from east to west. Everywhere defiles, rents, and torrent beds
are to be seen. The principal ridges from east to west are
Dhrun (5,177 feet), Gurangatti (3,906 feet), Tāloi (3,022 feet),
and Gokprosh, whose highest point is Janzāt (4,345 feet).
Gokprosh is famous as the scene of the defeat of the Baloch
rebels in 1898. Neither permanent inhabitants nor cultivated
lands exist. A few stunted trees and scrub jungle compose the
only vegetation. Sind ibex and mountain sheep are plentiful.

Hāmūn-i-Lora.—A depression in the Chāgai District of
Baluchistān, lying between 29° 8′ and 29° 37′ N. and 64° 44′
and 64° 59′ E. It is about 38 miles long, with an average
breadth of six miles. It receives the spill of the Pishīn Lora,
but, except after heavy rains, can be crossed in any direction.
The surface is white and impregnated with salt. Small rocky
hills rise to the south of it, the principal being Kaftār and
Gaukoh.

Siranda (25° 27′ to 25° 35′ N. and 66° 37′ to 66° 41′ E.).—A
lake in the Miāni *niābat* of the Las Bela State in Baluchistān.
It runs north and south, and when full is about 9 miles long by
2 miles broad. During the greater part of the year the average
depth is 3 to 5 feet, but the south-west corner, called Kun, is.

deeper. On the occurrence of floods the level of the water is raised 10 or 12 feet. The water is brackish, the lake having been formed by the gradual recession of the sea. Thousands of water-fowl resort to the lake in the cold weather, and it contains many small fish.

Hāmūn-i-Māshkel.—A large depression on the western frontier of Baluchistān, lying between 28° 2′ and 28° 30′ N. and 62° 36′ and 63° 27′ E. Its length from east to west is about 54 miles, and its breadth varies from 8 to 22 miles. It receives drainage from the south, east, and north, the principal supplies being from the Māshkel and Morjen rivers. There is never much water in it except for a short time after heavy rain. The greater part is covered with white saline efflorescence, and at Wād-i-Sultān is a small area containing good hard salt. On the north-west, and forming a separate basin, lies the Hāmūn-i-Tahlāb.

Zhob River.—A river in Baluchistān, rising on the east of Pishīn (30° 45′ N., 67° 33′ E.), and having a total length to its junction with the Gomal river of about 240 miles. Its chief affluents are the Toi or Kandīl river and the Sretoi from the north-west and west, and the Sāwar from the east. From the north it receives the drainage of the Toba-Kākar range, and from the south that of the hills dividing the Zhob valley from the catchment area of the Nāri river. The permanent perennial stream first appears at about 45 miles from the source, but, owing to the height of the scarped banks, the water cannot be utilized for irrigation till near Samakhwāl. Below this it is raised by artificial dams in several places.

Nāri.—A river in Baluchistān, also known as the Anambār and the Beji. It rises near Spīrarāgha and has a total length in Baluchistān of about 300 miles. The upper part of its course is known as the Loralai river, and after the junction of the latter with the Sehān it becomes the Anambār. On passing into the Marri country it is called the Beji. Near Bābar Kach it is met from the north-west by the Dādā and Sāngān streams, and shortly afterwards debouches into the Kāchhi plain, whence it branches into a number of channels (28° 30′ N., 67° 57′ E.), eventually reaching Sind. Its large catchment area covers the Loralai and Sibi Districts and Kachhi. The Nāri is subject to very heavy floods. Temporary embankments are erected in its bed to permit the cultivation of lands in the Loralai District, and a Government irrigation work to carry flood-water has also been constructed. All the permanent water-supply is used at Sibi for cultivation; and the central part of Kachhi entirely depends on its flood-water, which is raised by ingeniously contrived temporary earthen embankments of great size. Much of the summer flood-water, however, runs to waste.

Pishīn Lora[1].—A river in Baluchistān, having its source in the western slopes of the Kand mountain of the Toba-Kākar range and terminating in the Hāmūn-i-Lora. Its total length is about 250 miles. The principal affluents meet near Shādīzai in Pishīn. In addition to the Barshor Lora or main stream, they consist of the Kākar Lora, the Surkhāb, and the Shorarūd. Below the confluence of the upper tributaries the bed is 200 yards wide and lies between scarped banks about 20 feet high. The running stream, however, is usually not more than a few yards wide and quite shallow. On entering the hills west of Shorarūd the course becomes deep and narrow, until it debouches into the Shorāwak plain (30° 22' N., 66° 22' E.). Here it becomes dissipated into several channels which find their way through Nushki. The area drained includes the west of the Sarawān country, Quetta-Pishīn, and Nushki in Baluchistān, besides Shorāwak in Afghānistān. For purposes of irrigation water is taken off wherever it can be made available. The Shebo canal and Khushdil Khān reservoir in Pishīn are dependent on it for their supply; and in 1903 an embankment for irrigation purposes was constructed in the north of the Nushki *tahsīl* across the Bur channel.

Mūla.—A river in Baluchistān, rising in the Harboi hills and having a total length of 180 miles. As far as Kotra in Kachhi (28° 22' N., 67° 20' E.) it passes with a rapid fall through the Central Brāhui range; in its lower reaches many flats lie along its course. The upper course is known as the Soinda; a little lower it is called the Mishkbel, and from Pāshthakhān downwards it becomes the Mūla. Its principal affluents are the Malghawe, the Anjīra or Pissibel, and the Ledav. The Mūla drains the whole of the north-east of the Jhalawān country and also the south-west corner of Kachhi. Wherever possible, the small perennial supply of water is drawn off to irrigate the flats along the course of the river, while flood-water is utilized for cultivation in Kachhi. The Mūla Pass route to the Jhalawān country lies along its course.

Hab.—A river on the western frontier of Sind, Bombay Presidency, and for some distance the boundary between British territory and Baluchistān. It rises opposite the Porāli

---

[1] *Lora* is a Pashtū word signifying a channel carrying flood-water, as distinguished from *rūd*, a perennial stream.

river at the northern end of the Pab range, flows south-east for 25 miles, then due south for 50 miles, and finally south-west, till it falls into the Arabian Sea near Cape Monze, in 24° 54′ N. and 66° 42′ E., after a total length of about 240 miles. Except the Indus and the Gāj, it is the only permanent river in Sind. Its principal tributaries are the Sārūna, the Samotri, and the Wira Hab. As far as the Phusi Pass the course is confined and narrow. Thereafter it gradually widens, and for some 50 miles from its mouth is bordered by fine pasture land. Water is always to be found in pools, but the river is not utilized for irrigation.

Porāli.—A river in Baluchistān, draining the south of the Jhalawān country and the Las Bela State. It rises near Wad in 20° 33′ N. and 66° 23′ E., and enters the Pab range by a tortuous but picturesque channel. A course of 175 miles carries it to the sea at Miāni Hor. The principal affluents are the Kud, which drains the valley of Ornāch, the Tibbi, and the Lohendav. About five miles north of Sheh in Las Bela the Porāli bifurcates, and most of its flood-water is carried off by the Titiān, which enters the Siranda lake. Within the hills many flats are irrigated from this river, and the *niābat* of Welpat in Las Bela is also dependent on it. Temporary dams have been erected near Sheh and in the Titiān for purposes of cultivation. The Porāli has been identified with the ancient Arabis or Arabius.

Hingol.—The largest river in Baluchistān, 358 miles long. It rises at the head of the valley of Sūrāb, and drains the western side of the Jhalawān country and the north-eastern part of Makrān. The Hingol is known by a variety of names : as the Rej in Sūrāb, as the Gidar Dhor in the Gidar valley, and as the Nāl Kaur in the central part of its course. Its principal affluents join it in the south. They are the Mashkai, which meets the main stream under the name of the Pao, and the Arra. The Mashkai drains a very large area, including the Mashkai, Rāghai, and Gichk valleys. There is no continuous flow of water in the upper part of the course of the Hingol ; it meanders through stony valleys, the water being utilized wherever possible for cultivation, and constantly disappears in underground channels. From Kurragi in Jau downwards the supply becomes perennial. Thence the river passes through a series of magnificent but narrow gorges, and falls into the Arabian Sea in 25° 23′ N. and 65° 28′ E. Near its mouth is the celebrated shrine of Hinglāj.

Rakhshān.—A river in Baluchistān, rising near Shireza,

a point close to the eastern junction of the Central Makrān and Siāhān ranges. It traverses Panjgūr, on the west of which it is joined by the Gwārgo stream. It then turns northward, and joining the Mashkel river from Persia in 27° 10′ N. and 63° 27′ E., bursts through the Siāhān range by the fine defiles of Tank-i-Grawag and Tank-i-Zurrati, and runs under the latter name along the western side of Khārān to the Hāmūn-i-Māshkel. Its total length is 258 miles. Water from the Rakhshān is used for irrigation in Nāg-i-Kalāt, Panjgūr, and Dehgwar in Khārān.

Dasht.—A river in Baluchistān, draining the south-west part of Makrān, and formed by the junction of the Nihing and the Kech Kaur at Kaur-e-Awārān. The Kech Kaur, in its turn, is formed by the two streams, the Gish Kaur draining the Buleda valley, and the Kīl or Kūl Kaur which rises in the Central Makrān Range. From the source of the Kīl Kaur to the sea the course is 255 miles. The Nihing irrigates Mand and Tump, and the Kech Kaur waters Buleda, Sāmi, and Kech. The united streams then fertilize the Dasht valley, and fall into the sea by a creek which is tidal for 12 or 15 miles (25° 12′ N. and 61° 38′ E.). Irrigation is chiefly from flood-water or from pools in the bed. Much diluvion is caused by the floods of the Kech Kaur in Kech and Sāmi.

Zhob District.—A District situated in the north-eastern corner of Baluchistān, between 30° 32′ and 32° 4′ N. and 67° 27′ and 70° 3′ E., with an area of 9,626 square miles. Afghānistān and the Frontier Province lie to the north and north-west of the District ; its eastern boundary is marked by the Sulaimān range ; and the Loralai and Quetta-Pishīn Districts border it on the south and west respectively. The greater part of the District is covered with hills ; but it is intersected on the south by the great valley of the Zhob, and on the north by the smaller valley of the Kundar and its tributaries. The principal hill ranges are the SULAIMĀN on the east and the TOBA-KĀKAR range on the north, the latter being generally known as Kākar Khorāsān. It forms the great grazing-ground of the District and the summer quarters of many of the inhabitants of the Zhob valley. Both these ranges attain an altitude of nearly 11,000 feet above sea-level. On the south, a lower range separates the District from Loralai.

The valley of the ZHOB river itself is an alluvial plain, extending from Chari Mehtarzai, the watershed between the Zhob and Pishīn valleys, in the form of a crescent to the

Boundaries, configuration, and hill and river systems.

Gomal river, and contracting considerably near its north-eastern extremity. Numerous small valleys skirt it on either side, the most important being Haidarzai and Ismailzai. The Kundar rises from the central and highest point of the Toba-Kākar range, and runs first eastward and then north-eastward till it reaches the Gomal river near Domandi. In the upper part of its course it traverses a narrow valley, but from the point where it commences to form the boundary between British territory and Afghānistān, its bed is confined and steep. Among the hills lie numerous minor valleys, affording room for some scant cultivation. Occasionally wide open plains occur, such as that of Girdao.

The stratified rocks of the District include a varied series Geology. extending from the trias to the base of the pliocene. Intrusions of gabbro and serpentine, of Deccan trap age, and extensive accumulations of sub-recent and recent deposits also occur. The most interesting feature is the continuation of the Great Boundary Fault of the Himālayas, north of the Zhob valley. The serpentine intrusions contain valuable deposits of chrome-iron ore and some asbestos of marketable value.

No scientific information is available regarding the vegetation Botany. of Zhob. The principal trees found in the highlands are olive, pistachio, and edible pine. Tamarisk borders the beds of the streams, and acacia occurs in the lower hills. Southernwood (*Artemisia*) scrub abounds in the uplands, and *Haloxylon salicornicum* in the lower valleys.

Wolves and foxes are common. Mountain sheep and Fauna. straight-horned *mārkhor* are found in the higher hills. *Chikor* and *sīsī* are plentiful. Some wild hog are to be seen along the Zhob river. Fishing is obtainable in the Zhob and Sāwar streams.

The climate is dry and, in the winter months, bracing. Climate, Dust-storms occur in summer from July to September, accom- temperature, and panied by thunderstorms, rendering the climate of Lower rainfall. Zhob somewhat enervating. On the other hand, Upper Zhob and the highlands possess excellent climatic conditions. The temperature varies with the height above sea-level; in Lower Zhob, the heat is unpleasant from May to September, but the western end of Upper Zhob is cool in summer and cold in winter. In many places the water-supply is brackish. The rainfall is scanty and variable, the annual average at Fort Sandeman being about 10 inches, but it is less in Upper Zhob. The summer rainfall is received in July and August; in winter, March is the rainiest month. At this time of year some snow falls over the greater part of the District.

History and archaeology.

As the cradle of the Afghān race, Zhob possesses a peculiar historical interest. Hiuen Tsiang describes the Afghāns as living here in the seventh century, and hence they emerged to seek riches and even empire in India. In 1398 we read of an expedition led by Pīr Muhammad, grandson of Tīmūr Lang, against the Afghāns of the Sulaimān Mountains, which appears to have penetrated this District. The migration of the Yūsufzais from Zhob to Swāt has been recently traced. We can, however, only catch occasional glimpses of the ancient history of the country. In the middle of the eighteenth century Ahmad Shāh, Durrāni, conferred the title of Ruler of Zhob on Bekar Nika, the head of the Jogizai family of the Zhob Kākars; and this family continued to exercise authority over the Kākars until the British were first brought into contact with them. The most notable men at this time were Shāh Jahān, and his cousin and rival, Dost Muhammad. Though worsted in a fight at Baghao, near Sanjāwi, in 1879, Shāh Jahān continued to commit outrages in British territory; and a small military expedition was undertaken against the Zhob Kākars in 1884, upon which most of the headmen submitted. Shāh Jahān fled; but the Kākars continued to give trouble until, in 1888, Sir Robert Sandeman marched to Mīna Bāzār, after which, at the invitation of the chief of the Mando Khels, he visited Apozai, where the Mando Khels asked to be taken under British protection and offered to pay revenue. A similar request was preferred a little later by Shāh Jahān himself. The offer of the people was accepted, a Political Agent was appointed, and a small garrison was placed at Fort Sandeman in 1889. The Bori, Sanjāwi, and Bārkhān *tahsīls*, which had hitherto formed part of the Thal-Chotiāli District, were included in Zhob, but Sanjāwi was re-transferred to Thal-Chotiāli in 1891 and Bārkhān in 1892. All of them have been included in the Loralai District since 1903.

Many old forts and mounds, the construction of which is attributed by the people to the Mughals, are scattered throughout the country, but they have not yet been archaeologically examined. An interesting discovery of punch-marked coins, some of which possibly date so far back as 400 B.C., has been made in the Shirāni country.

The people, their tribes and occupations.

The District possesses 245 villages. The only place of importance is FORT SANDEMAN town. The population of the entire District was enumerated for the first time in 1901, and was found to number 69,718, or seven persons per square mile. The following table gives statistics of area, &c., by *tahsīls* according to the Census of 1901 :—

Tahsil.	Area in square miles.	Number of		Population.	Population per square mile.
		Towns.	Villages.		
Kila Saifulla . .	2,768	...	60	19,229	7
Hindubāgh . .	3,275	...	76	15,777	5
Fort Sandeman .	3,583	1	109	34,712	11
Total	9,626	1	245	69,718	7

The indigenous inhabitants are all Afghāns, who numbered 63,000 in 1901. The Kākars constitute the principal tribe of the District, numbering 43,000 ; the majority (38,000) belong to the great clan known as Sanzar Khel. Other Afghān tribes are the Ghilzais (7,500), who are mostly Powindah Nāsirs and Sulaimān Khels, resorting to the District in the course of their annual migrations for purposes of trade and pasture ; the Mando Khels (4,300) ; and the Shirānis (7,000). The Bargha, or upper Shirānis, are alone subject to the Zhob authorities, the Largha, or lower Shirānis, living within the limits of the Frontier Province. By religion the people are Muhammadans of the Sunni sect. They speak the southern dialect of Pashtū. The majority are cultivators, but a few live by flock-owning, and many supplement their means of livelihood by labour and engaging in transport. Some, especially the Bābars, annually visit the Chāgai hills and parts of Afghānistān to collect asafoetida. Ancestor worship is much in vogue, and the tomb of Sanzar Nika, the progenitor of the Sanzar Khels, which lies 27 miles from Fort Sandeman, is held in great reverence. The Jogizais, too, are endowed with theocratic attributes in the eyes of their fellow tribesmen.

The soil of the District is largely mixed with gravel, especially Agricul-in the hill tracts. In Central Zhob it is alluvial and sometimes ture. sandy. The Fort Sandeman *tahsīl* possesses a productive red clay. Land is generally allowed to lie fallow for two to four years after cultivation. Two harvests are obtained : the spring crop (*dobe*), sown from October to January and reaped in May and June ; and the autumn crop (*mane*), sown in June and July and reaped in November and December. The principal harvest is in spring, and the most important food-grain is wheat, which is estimated to cover about 63 per cent.

of the area under crops. The autumn crop consists of maize, millets, and rice. Gardens are numerous in Upper Zhob, and are increasing in number ; they contain mulberries, apricots, pomegranates, and grapes. Melons and lucerne are also grown. The cultivators are chiefly peasant proprietors and their holdings are minute. They are indifferent husbandmen, but British rule has resulted in a considerable extension of cultivation. The advances made to the people during the seven years ending in 1904 amounted to Rs. 44,800 for land improvement and Rs. 29,800 as agricultural loans.

Cattle, horses, sheep, &c.   Unlike other parts of Baluchistān, the District possesses few good horses. The breed of bullocks is small ; camels of good quality are numerous at certain times of the year. Sheep and goats are owned in large numbers. Considerable flocks and herds are brought into the District by Sulaimān Khel Ghilzais and other nomads.

Irrigation.   It is estimated that about 74 per cent. of the area under cultivation is irrigated. The 'dry' crop area is probably capable of considerable extension. The sources of irrigation are springs, *kārez* or underground channels, and streams. Of the latter, the principal are the Zhob river, Saliāza, Sāwar, Rod Fakīrzai, and Kamchughai. Dams are thrown across their beds and the water is led in open channels to the cultivable land. Some of these channels exhibit considerable ingenuity of construction, but they require constant repair.

Forests.   Measures for forest conservancy have been taken since 1894 under the executive orders of the Political Agent. No areas have been 'reserved,' but the felling of green forest trees is prohibited throughout the District. The principal trees in order of frequency are olive, two kinds of pistachio, edible pine, and ash. Proposals for the 'reservation' of the pine and olive forests of Shīnghar, Kapīp, and other parts are under consideration. From 1894 to 1900 the forest revenue, chiefly on account of royalty on wood, averaged Rs. 2,600 per annum and the expenditure Rs. 953. In 1903–4 the revenue was Rs. 3,600 and the expenditure Rs. 1,300.

Minerals.   Traces of coal have been noticed in Central Zhob, and asbestos exists in the Hindubāgh *tahsīl*. Earth-salt is collected in Central Zhob from the Multani tracts, and also in Kākar Khorāsān by the Lawānas.

Arts, manufactures, and commerce.   As might be expected in so backward a country, the manufactures are of the most primitive kind. Felts and felt coats are made by the women for home requirements. Woollen carpets and bags of various sorts are also turned out by weavers, who ply their profession from house to house, but there is no trade in these articles. The chief exports are wool and *ghī*, while imports include sugar, rice, pulses, metals, piece-goods, leather-ware, and salt. Trade is chiefly carried on with Dera Ismail Khān District in the Frontier Province. A small quantity of earth-salt is exported to Afghānistān.

The total length of partially metalled roads is $27\frac{1}{4}$ miles, and Roads. of unmetalled tracks, 766 miles. The southern and north-eastern parts are well provided with means of communication, but not the north-western portions. Of the two main roads, the best runs from the Harnai railway station through Loralai to Fort Sandeman, a distance of 168 miles, the first 22 of which lie in the Sibi and 76 in the Loralai District ; the other starts from the Khānai railway station and traverses the Zhob valley to Fort Sandeman, a distance of 168 miles, the first 45 of which lie in the Quetta-Pishīn District. A road has been constructed through the Dhāna Sar Pass to Dera Ismail Khān, 115 miles in length, of which about $47\frac{1}{2}$ lie in Zhob. A survey for a railway line from Khānai in Quetta-Pishīn to Dera Ismail Khān District through the Zhob valley was carried out in 1890–1.

The succession of three dry years, which culminated in Famine. 1900–1, severely affected the District, and great numbers of cattle died. About Rs. 57,000 was spent on relief works in Central and Lower Zhob, and Rs. 3,825 was advanced for the purchase of seed-grain and bullocks. Grazing tax amounting to Rs. 18,800 was suspended. The Government revenue, which is taken in kind, decreased largely.

The District is part of the Agency Territories and forms the District charge of a Political Agent. It is divided into two subdivisions, subdivi-Fort Sandeman and Upper Zhob. An Assistant Political Agent staff. holds charge of the Fort Sandeman subdivision, with an Extra Assistant Commissioner to assist him. Another Extra Assistant Commissioner is in charge of the Upper Zhob subdivision. Each of the three *tahsīls*, Fort Sandeman, Kila Saifulla, and Hindu-bāgh, has a *tahsīldār* and a *naib-tahsīldār*, with the exception of the first, which has two *naib-tahsīldārs*. All these officers Civil jus-exercise civil and criminal powers, and the Assistant Political tice and Agent and Extra Assistant Commissioners are empowered to crime. hear appeals from *tahsīldārs* and *naib-tahsīldārs*. The Assistant Political Agent is also a Justice of the Peace. The Political Agent is the District and Sessions Judge. The number of cognizable cases reported in 1903 was ten, convictions being obtained in six cases. Only one case of murder occurred. The number of criminal and civil cases disposed of by the courts in 1903-4 was 38 and 253 respectively. As a rule, however, cases in which the people of the country are concerned are referred to councils of elders (*jirgas*) for an award according to tribal custom under the Frontier Crimes Regulation. The number of cases thus disposed of in 1903-4 was 930, which included 10 murder cases and 11 of adultery.

Land revenue administration.

In pre-British times the tribes paid no land revenue, except an occasional sheep or a felt to their chief. Since the British occupation revenue has been realized at a uniform rate of one-sixth of the produce, the Government share being determined either by the appraisement of standing crops or by actual division. In some villages, which have permanent sources of irrigation, summary cash assessments, based on the realizations of the preceding three or five years, have been introduced as a temporary measure. A survey and record-of-rights have been prepared for the Upper Zhob subdivision. Grazing tax is also levied, and forms a large item in the total revenue. A special arrangement was made in 1897-8 with the Punjab Government (now the Frontier Province) for the collection of grazing tax from the Sulaimān Khel Ghilzais, who pasture their flocks in both Zhob and the Derajāt, by which three-fifths of the net receipts were paid to the latter Government up to 1902. The receipts from land revenue, including the total grazing tax of the District and royalty levied on firewood in the Fort Sandeman *tahsīl*, amounted to 1·1 lakhs in 1903-4, of which Rs. 2,513 was realized by summary assessments. The grazing tax produced Rs. 31,000. The total revenue from all sources was 1·3 lakhs. The Jogizais, to whom reference has already been made, and other leading families enjoy revenue-free grants, grain allowances, and exemption from grazing tax. The total annual value of these concessions is estimated at Rs. 8,800.

Army.

The regular troops at Fort Sandeman furnish detachments to the outposts at Khān Muhammad Kot in the Loralai District, and to Mīr Alī Khel, Mānikwā, and Wazīr Bāgh in Zhob. The detachments at Hindubāgh and Kila Saifulla are furnished from Loralai. The Zhob Levy Corps guards a long line of frontier from the Gomal to Toba in sixteen outposts. It consists of six companies of infantry, aggregating 632 men, and four squadrons of cavalry of 423 men under three European officers, with its head-quarters at Fort Sandeman.

Police and jails.

An Honorary Assistant District Superintendent held charge of the police, temporarily, in 1903-4. It is proposed that a permanent officer should be appointed. In 1904 the force numbered 199 men, of whom 8 were inspectors and deputy-inspectors, and 32 were horsemen. They are distributed in five stations. Much of the work which ordinarily falls on the police in India is performed by local levies. They numbered (1904) 304, including 179 horsemen, of whom 50 men were employed on postal service. The three subsidiary jails accommodate 106 male and 19 female prisoners. Ordinarily, prisoners whose sentences do not exceed six months are retained in the local jails, while long-term prisoners are sent to Quetta and Shikārpur.

The District possessed four primary schools in 1904, includ- Education. ing one girls' school. The total number of pupils was seventy-four, including twenty-four girls, and the cost was Rs. 1,688, of which Rs. 970 was paid from general revenues, Rs. 324 realized from fees and contributions, and the balance paid from Local funds. The people care little for education, though *mullās* impart some elementary instruction in mosques. The number of pupils receiving instruction in such schools is estimated at about 350.

There are four dispensaries in the District, including one Medical. for females at Fort Sandeman, which is aided by the Lady Dufferin Fund. They had accommodation for thirty-six in-patients in 1903. The average daily attendance of such patients was 18, and the total attendance of all patients was 30,432. The expenditure amounted to Rs. 9,400, the greater part of which was provided from Provincial revenues. Malarial fever is the most prevalent disease, followed by diseases of the digestive organs. Owing to the cold, rheumatism is common. Resort is had to inoculation by the people living at a distance from the *tahsīl* head-quarters. Vaccination Vaccina-is gradually making headway, but is resorted to only when out- tion. breaks of small-pox occur. The number of persons successfully vaccinated in 1903 was 43 per thousand.

[*Administration Report of the Baluchistān Agency for* 1890–1.]

Upper Zhob.—A subdivision of the Zhob District, Baluchistān, consisting of the HINDUBĀGH and KILA SAIFULLA *tahsīls*.

Fort Sandeman Subdivision.—A subdivision and *tahsīl* of the Zhob District, Baluchistān, forming the north-eastern corner of the District, and lying between 30° 39' and 32° 4' N. and 68° 58' and 70° 3' E., with an area of 3,583 square miles. Population (1901), 34,712. The land revenue, including grazing tax and royalty levied on wood, amounted in 1903–4 to

Rs. 40,000. The head-quarters station is FORT SANDEMAN (population, 3,552). The *tahsīl* possesses 190 villages. The country is hilly, and intersected by the valley of the Zhob and many minor valleys. Cultivation is sparse and backward. The Girdao plain is covered with rich pasture in years of good rainfall. The Shīnghar spurs of the Sulaimān range contain fine forests of edible pine.

Kila Saifulla.—A *tahsīl* of the Upper Zhob subdivision of the Zhob District, Baluchistān, situated between 30° 32′ and 31° 43′ N. and 68° 9′ and 69° 18′ E. It lies along the central part of the valley of the Zhob river, and also includes part of the Toba-Kākar range known as Kākar Khorāsān. Its area is 2,768 square miles, and population (1901) 19,229. The land revenue, including grazing tax, amounted in 1903–4 to Rs. 44,000. The head-quarters station is Kila Saifulla, and the *tahsīl* contains sixty villages. The majority of the people are Sanzar Khel Kākars, who combine flock-owning with agriculture. They cultivate considerable rain-crop areas. The Jogizais, once the ruling family in Zhob, live in this *tahsīl*. Earth-salt is manufactured, and traces of coal have been found. A small trade is done in fox-skins.

Hindubāgh.—A *tahsīl* of the Upper Zhob subdivision of the Zhob District, Baluchistān, lying between 30° 36′ and 31° 50′ N. and 67° 27′ and 68° 46′ E. It is bounded on the north by the Toba-Kākar range, which separates it from Afghānistān. Its area is 3,275 square miles, and population (1901) 15,777. The land revenue, including grazing tax, amounted in 1903–4 to Rs. 24,000. The head-quarters station, which bears the same name as the *tahsīl*, lies in the south-west corner. The *tahsīl* possesses seventy-six villages. The main valley, called Zhob *nāwah* from its boat-like shape, lies along the upper course of the Zhob river, while the northern part covers the grassy uplands of Kākar Khorāsān. The greater part of the cultivation is irrigated; rain crops are comparatively insignificant. Under the Kand mountain lies the picturesque glen of Kamchughai. Asbestos deposits exist in the valley.

Fort Sandeman Town.—Head-quarters station of the Zhob District, Baluchistān, situated in 31° 21′ N. and 69° 27′ E. It was first occupied in December, 1889. To the natives the locality is known as Apozai; it received its present name from its founder, Sir Robert Sandeman. The station stands about 6½ miles east of the Zhob river, in an open plain 4,700 feet above sea-level. A ridge rises 150 feet above the surface of the plain to the north, on which stands the residence of the Political Agent, known as the Castle. The military lines, bazar, dispensaries, and school lie below. The nearest railway station in Baluchistān is Harnai, 168 miles; Bhakkar, the railway station for Dera Ismail Khān, is 122 miles. The population numbered 3,552 in 1901. The garrison includes a native cavalry and a native infantry regiment, and Fort Sandeman is also the head-quarters of the Zhob Levy Corps. A supply of water for drinking purposes, carried by a pipe nine miles from the Saliāza valley, was inaugurated in 1894, at a cost of a little over a lakh of rupees. Water for irrigation is also obtained from the same source, and by this means many fruit and other trees have been planted. A Local fund has existed since 1890; the income during 1903-4 was Rs. 18,000 and the expenditure Rs. 17,000. One-third of the net receipts from octroi is paid over to the military authorities. A small sanitarium, about 8,500 ft. above sea-level, exists about 30 miles away at Shīnghar on the Sulaimān range, to which resort is made in the summer months.

Loralai District.—A District of Baluchistān, lying between 29° 37′ and 31° 27′ N. and 67° 43′ and 70° 18′ E., with an area of 7,999 square miles. It derives its name from the Loralai stream, an affluent of the Anambār or NĀRI. On the north it is bounded by the Zhob District; on the east by the Dera Ghāzi Khān District of the Punjab; on the south by the Marri country; and on the west by the Sibi District. It consists of a series of long but narrow valleys hemmed in by rugged mountains, which vary in elevation from 3,000 to 10,000 feet. Those occupying the west and centre have a direction from east to west, and form the upper catchment area of the Anambār river. Those on the east run north and south, and their drainage bursts through the Sulaimān range into the Indus valley. The western ranges, which are highest, contain much juniper and some fine scenery. The central hills consist of three parallel ranges stretching out to meet the Sulaimān range. They are known locally as the Dāmāngarh on the north; the Krū and Gadabār hills in the centre; and the Dabbar range with its eastern continuation on the south. The Anambār, which debouches into Kachhi under the name of the Nāri, is the principal river of the District. It is formed by the junction of the Loralai, Mara, and Sehān streams, and is joined lower down by the Lākhi and Narechi. On the east three rivers carry the drainage into the Indus valley: the Vihowa and Sanghar in the Mūsā Khel *tahsīl* and the Kāhā

[margin note: Boundaries, configuration, and hill and river systems.]

in Bārkhān. Each of these has a small perennial flow, which frequently disappears, however, beneath the stony bed.

Geology. The strata exposed in Loralai include the upper, middle, and lower Siwāliks (upper and middle miocene) ; the Spīntangi limestone and Ghāzij group (middle eocene) ; volcanic agglomerates and ash-beds of the Deccan trap ; the Dunghān group (upper cretaceous) ; belemnite beds (neocomian) ; and some inliers of the massive limestone (jurassic). A triassic belt occurs between the District and the Zhob valley.

Botany. The District is barely clothed with vegetation. The trees include juniper and pistachio at the higher levels, and acacia and olive on the lower hills. The poplar (*Populus euphratica*) and willow are also to be found. Tamarisk, the wild caper, and dwarf palm occur in the valleys. Myrtle groves are found in the Smāllan valley, and box on the summit of the Mūsā Khel hills. *Shīsham* (*Dalbergia Sissoo*) has been introduced at Duki and grows well. Orchards are numerous in Sanjāwi, Duki, and Bori, containing apricots, mulberries, and pomegranates. Vineyards are also common in Sanjāwi, Thal, and round Loralai town. Grapes in Thal sell for R. 1 per donkey-load in the season.

Fauna. Game is not abundant. Some mountain-sheep and *mārkhor* are to be found in the hills, while leopards, black bears, and numerous wolves and hyenas also occur. A few wild hog are occasionally met with. Snakes are numerous. Fishing is to be had in the Anambār and larger streams.

Climate, temperature, and rainfall. The climate varies with the elevation, but on the whole is dry and healthy. In the west of the District the seasons are well marked ; the summer is cool and pleasant, but the winter is intensely cold, with hard frosts and falls of snow. In the south and east the temperature is more uniform, but the heat in summer is great. The Bori valley is subject to high winds, which are very cold in early spring and have been known to cause considerable mortality. The rainfall is light, the annual average being about seven inches. In the western parts both summer and winter rain and also some snow are received. The rest of the District depends chiefly on the summer rainfall, which everywhere exceeds that in winter.

History and archaeology. The District in ancient times formed the most eastern dependency of the province of Kandahār, and, like that place, its possession alternated from time to time between Mughal, Safavid, and Afghān. Its capital was Duki, which was generally garrisoned. The District lay across one of the main routes from

India to Western Asia via the Sakhi Sarwar Pass and Pishīn. It provided a contingent of 500 horse and 1,000 foot for Akbar, besides other contributions. In 1653, when Dārā Shikoh, son of the emperor Shāh Jahān, advanced against Kandahār from India, he occupied Duki, which had been held by Persian troops. Later, the District passed to the Durrānis and their successors.

The steps by which different parts of the District have come under British control were gradual. In 1879 the Duki *tahsīl* was ceded under the Treaty of Gandamak and a force under General Biddulph was sent with Sir Robert Sandeman to explore the country, in the course of which a successful engagement was fought with the Zhob and Bori Kākars at Baghao. The country had long been the battle and raiding-ground of rival tribal factions, the Marris fighting the Lūnis and Khetrāns, the Tarīns of Duki being at constant war with the Dumars and Utmān Khels, and the Mūsā Khels raiding the Baloch of the Punjab. In consequence, the inhabitants of Sanjāwi were brought under British protection in 1881. In 1879 a detachment had been stationed at Vitākri in the Khetrān country to check the Marris, but it was shortly afterwards withdrawn, and in 1887 the valley (now the Bārkhān *tahsīl*) was also taken under protection. In 1884 frequent raids by the Kākars from the north culminated in an attack on coolies employed at Duki, which led to a small punitive expedition being dispatched under General Tanner. The tribesmen submitted, and the expedition eventually resulted in the occupation of the Bori valley in 1886. A settlement had been made with the Mūsā Khels after the expedition of 1884; and on the occupation of Zhob in 1889 the Mūsā Khel country was included in that Agency, a *tahsīl* being established there in 1892. The District, as it now exists, was formed in 1903, the Mūsā Khel and Bori *tahsīls* being transferred from the Zhob District and the Duki, Sanjāwi, and Bārkhān *tahsīls* from the Thal-Chotiāli District.

Interesting mounds and ruins mark the course of the ancient trade route from India to Central Asia, but they have never been explored. Remains of large dams, probably used as water reservoirs, exist here and there. A find of coins of the Caliph Marwān II (A.D. 745) has been made at Dabbar Kot.

The District has 400 inhabited villages and a population (1901) of 67,864, or eight persons per square mile. Of these about 95 per cent. are Muhammadans of the Sunni sect, and *The people, their tribes and occupations.*

most of the remainder Hindus. The following table gives
statistics of area, &c., by *tahsīls* in 1901 :—

Tahsī.	Area in square miles.	Number of		Population.	Population per square mile.
		Towns.	Villages.		
Mūsā Khel . .	2,213	...	55	15,537	7
Bārkhān . .	1,317	...	114	14,922	5
Duki . . .	1,951	...	66	12,365	6
Sanjāwi sub-*tahsīl*	446	...	37	6,866	15
Bori . . .	2,072	1	128	18,174	9
Total	7,999	1	400	67,864	8

The principal inhabitants are the Kākars (18,400), Khetrāns
(13,600), Mūsā Khels (10,500), Dumars (5,300), Tarīns (3,400),
Lūnis (2,600), and Pechi Saiyids (800). With the exception
of the Khetrāns, they are all Afghāns. The Khetrāns claim
both Baloch and Afghān affinities, but the majority of them
are probably of Jat extraction. Most of the people are small
cultivating proprietors, but some sections are almost entirely
flock-owners. A small number of weavers live in the Bārkhān
tahsīl. Hindus carry on the retail trade of the District. In
Duki a curious instance of the assimilation by Hindus of
Muhammadan traits and dress is to be found in the Rāmzais,
who have long shared the fortunes of the Hasni section of the
Khetrāns. They dress as Baloch and are expert swordsmen
and riders. The language of the Afghāns is Pashtū, and that
of the Khetrāns is a dialect akin to Western Punjābi.

Agricul-
ture.

The soil of the Bori valley consists of a reddish loam, and is
highly productive if properly cultivated. In Duki a pale grey
loess occurs; elsewhere extensive gravel deposits are to be
found, mixed occasionally with tracts of good cultivable red
clay. The principal harvest is the *dobe*, or spring crop, which
is sown after the autumn rains, and matures with the aid of the
winter moisture. The *mane*, or autumn crop, is sown on the
summer rainfall. Lands which depend on rain or floods are
generally cropped each year, if the rainfall is sufficient; irrigated
lands are allowed to lie fallow for one to four years, but lands
close to villages which can be manured are sometimes cropped
twice a year.

Of the five *tahsīls*, Sanjāwi alone has been brought under
settlement. Its total cultivable area is 9,700 acres, of which
7,600 acres are cultivated. The irrigated area represents 74
per cent. of the whole, but in other *tahsīls* the 'wet' crop area is

not so large. A record-of-rights has been made in the Bori
tahsīl. The principal crops are wheat, *jowār*, maize, and rice,
with a small amount of tobacco grown in Sanjāwi. There has
been a considerable extension of fruit gardens, and quantities of
grapes, apricots, and pomegranates are produced. Melons of
a superior kind, vegetables, and lucerne are cultivated in some
parts. With the introduction of peace and security, the
inhabitants are settling down to cultivation, which is gradually
extending. Under an arrangement made in 1897, the Leghāri
chief is bringing lands in Bārkhān under cultivation, which had
lain waste for ages owing to their exposure to Marri marauders.
Most of the cultivators readily avail themselves of Government
loans, but a few have religious scruples about paying interest.
Between 1897-8 and 1903-4, Rs. 83,700 was advanced for
agricultural improvements, and Rs. 56,000 for the purchase of
seed and bullocks.

Much of the wealth of the District consists in its herds of Cattle,
cattle, sheep, and goats, which find ample grazing in the plains horses,
of Saharā in Mūsā Khel, of Ranrkan in Bārkhān, of Thal and sheep, &c.
Chamālang in Duki, and round the base of Akhbarg in the
Dumar country. The Mūsā Khels possess a comparatively
larger number of camels and donkeys, while Bārkhān, Sanjāwi,
and Bori have more sheep and goats. The Buzdār breed of
donkeys is excellent. Some of the best horses in the country
were to be had in Duki and Bārkhān in the early days of the
British occupation ; but many of the mares were bought up, and
the breed has somewhat deteriorated. Government stallions
are stationed at Loralai, Duki, and Bārkhān. The branded
mares numbered seventy-three in 1904. Many camels, sheep,
and goats are brought into the District by nomads at certain
times of the year.

The principal sources of irrigation are streams, springs, and Irrigation.
kārez. The Persian wheel is used in the Bārkhān *tahsīl.* Of
the 475 revenue villages, 173 have permanent irrigation, 111
are partially irrigated, and 191 depend entirely on rainfall and
floods. Besides the Sanjāwi *tahsīl,* the chief irrigated areas
are the Duki and Lūni circles in the Duki *tahsīl,* and the
Nāhar Kot circle in the Bārkhān *tahsīl.* The Bori *tahsīl*
possesses the largest number of *kārez,* sixty-two. A flood-
water channel was constructed by Government in 1903, at
a cost of Rs. 40,000, to take off the flood-water of the
Anambār to irrigate lands in the Thal plain.

The District contains about 55 square miles of state forests. Forests.
They include two tracts of juniper forest, covering 4,000 acres,

in the Sanjāwi *tahsīl*; the Gadabār forest, chiefly *Acacia modesta*, of about 35 square miles on the Gadabār hill; and Sūrghund and Nargasi, with an area of 14 square miles, the former containing *Prunus eburnea* and some juniper, and the latter pistachio. Timber fellings are also regulated in the Kohār pistachio forest. A grass Reserve is maintained on the banks of the Narechi.

Minerals. A coal seam in the Chamālang valley in Duki was examined by an officer of the Geological Survey in 1874, but he held out no hopes of a workable thickness being obtained. Traces of coal have also been noticed in the Sembar hills in the same *tahsīl*, and an extensive seam occurs within 1½ miles of Duki village.

Arts, manufactures, and commerce. Felts and felt coats (*khosā*), which are in daily use, are made by the women of the country. Mats and many other articles are woven from the dwarf-palm (*pīsh* or *dhorā*), also for domestic purposes. Bārkhān possesses an industry in the manufacture of carpets, saddle-bags, nose-bags, &c., in the *darī* stitch, which were once much admired; but the use of cheap aniline dyes has injuriously affected the trade, and the products are now inferior in quality. The articles are sold locally and also exported to the Punjab. The District produces grain, *ghī*, and wool, of which the last two are exported. Trade is carried on either with Sind through Harnai, or with the Dera Ghāzi Khān District. Transit dues, which formerly caused much hindrance to trade, have now been abolished.

Communications. The District possesses two excellent roads which intersect near Loralai: the Harnai-Fort Sandeman road, of which 76 miles lie in the District; and the Pishīn-Dera Ghāzi Khān road, of which 175 miles are within the District boundaries. The total length of communications consists of 288 miles of metalled or partially metalled roads, and 737 miles of unmetalled tracks and paths.

Famine. Bori and Sanjāwi are best protected from famine. The Mūsā Khel and Bārkhān *tahsīls* and large portions of the Duki *tahsīl* depend on the rainfall for their cultivation, and are severely affected by its local failure. The Bārkhān *tahsīl* suffered from drought in 1840, in 1860, and also in 1883. Scarcity again occurred in 1897–8; and Rs. 20,000 from the Indian Famine Relief Fund was spent in the Duki and the Bārkhān *tahsīls* during that year and in 1900–1, chiefly in providing cattle and seed-grain. The grazing tax was remitted and advances were made in other parts. The drought continued up to 1900, and revenue to the amount of Rs. 23,700 was suspended, nearly half of which represented the grazing tax of the Mūsā Khel *tahsīl*. Further advances were made in the following year to enable the people to recover. Between 1899 and 1901 about Rs. 69,000 was also spent on relief works from Imperial revenues.

The District is composed of two units, officially known as the Duki and Loralai Districts, the former belonging to British Baluchistān and the latter to Agency Territories. For purposes of administration it is treated as the single charge of a Political Agent, who is also Deputy-Commissioner. It is divided into three subdivisions: Bori, which is in charge of an Assistant Political Agent, and Mūsā Khel-Bārkhān and Duki, each of which is in charge of an Extra Assistant Commissioner. A *tahsīldār* and a *naib-tahsīldār* are posted to each *tahsīl* except Sanjāwi, which is in charge of a *naib-tahsīldār* who has the powers of a *tahsīldār*. District subdivisions and staff.

The *naib-tahsīldārs* and *tahsīldārs* exercise both civil and criminal jurisdiction, appeals from their decisions lying to the officers in charge of subdivisions. The Political Agent and Deputy-Commissioner is the principal Civil and Sessions Judge. In 1903–4 the civil suits decided in the District numbered 273 and the criminal cases 62. Cases in which the people of the country are concerned are generally referred to councils of elders (*jirgas*) for an award according to tribal custom under the Frontier Crimes Regulation, the final order being passed by the Political Agent. Such cases numbered 1,652 in 1903–4, including 14 cases of murder, 32 of adultery, and 13 of adultery with murder, as well as 71 inter-Provincial cases with the Punjab. Civil justice and crime.

In the time of Akbar the territory of Duki paid Rs. 120 in money, 1,800 *kharwārs* of grain, 12,000 sheep, and 15 Baluchi horses, besides contributing a military contingent; and under Afghan rule the same system appears to have been continued. Since the submission of the tribes to British rule, the revenue has been levied at a uniform rate of one-sixth of the gross produce. The Government share is generally determined by appraisement. The small *tahsīl* of Sanjāwi is under a fixed cash assessment for a period of ten years from 1901. Grazing tax is collected either by actual enumeration of the animals or in a lump sum fixed annually. In Mūsā Khel it forms the largest part of the revenue. It amounted for the whole District in 1903–4 to Rs. 28,600. Land revenue administration.

The receipts from land revenue and grazing tax, including the royalty levied on firewood, amounted in 1903–4 to

2·1 lakhs, which gives an incidence of Rs. 3 per head of population. The revenue from all sources during the same year was 2·3 lakhs.

Local boards.
In addition to the Loralai bazar fund, two other Local funds are maintained, the money raised being spent on sanitary establishments and watch and ward. The receipts in 1903–4 were Rs. 12,300 and the expenditure Rs. 15,100.

Army, police, and jails.
The regiments at Loralai furnish cavalry detachments at Gumbaz and Murgha Kibzai, and infantry guards for the sub-treasuries at Hindubāgh and Kila Saifulla in the Zhob District. Detachments from the cavalry regiment at Fort Sandeman are located at Mūsā Khel and Khān Muhammad Kot, and a small infantry detachment is stationed at Drug. Owing to the recent formation of the District, the police arrangements are still in a state of solution. The police force is at present directly controlled by the police officer at Fort Sandeman. In 1904 it consisted of 5 deputy inspectors and 112 men, including 14 horsemen, and held five posts. The levies numbered 392 men, of whom 7 were headmen and 224 mounted men. They were distributed in twenty-eight posts, and included seventy-four men employed on postal and telegraph service. The number of subsidiary jails or lock-ups was five, with accommodation for 125 male and 20 female prisoners. Convicts whose term exceeds six months are generally sent to Shikārpur in Sind.

Education.
In 1904 the number of primary schools was five, with eighty-four boys; the total cost was Rs. 1,800, of which Rs. 951 was paid from Provincial revenues, the balance being met from fees and Local funds. Elementary instruction, chiefly of a religious character, is given to about 850 boys and 180 girls in mosque schools, the largest number being in the Bārkhān *tahsīl.*

Medical.
Each of the five *tahsīls* possesses a dispensary. That at Sanjāwi is moved, during the summer months, to Ziārat in the Sibi District. There is accommodation for twenty in-patients. The average daily attendance of such patients in 1903 was 19, and the total average daily attendance of all patients 208. The cost of the dispensaries was about Rs. 10,000, which was wholly met from Provincial revenues.

Vaccination.
Vaccination is optional and the majority of the people still resort to inoculation. It is only when small-pox breaks out that the services of the Government vaccinators are requisitioned. Statistics of vaccination are not available.

['Report on the Geology of Thal-Chotiāli and a part of the Marri Country,' *Records of the Geological Survey of India,* vol.

xxv, part 1.—Surgeon-Major O. T. Duke: *An Historical and Descriptive Report of the Districts of Thal-Chotiāli and Harnai.* (Calcutta, 1883.)]

Mūsā Khel-Bārkhān.—A subdivision of the Loralai District, Baluchistān, comprising the *tahsīls* bearing the same names.

Duki Subdivision.—A subdivision of the Loralai District, Baluchistān, comprising the *tahsīls* of Duki and Sanjāwi.

Mūsā Khel.—A *tahsīl* of the Mūsā Khel-Bārkhān subdivision, in the north-eastern corner of the Loralai District, Baluchistān, situated between 30° 17′ and 31° 28′ N. and 69° 28′ and 70° 15′ E. Its area is 2,213 square miles, and population (1901) 15,537; the land revenue in 1903–4 amounted to Rs. 24,000. The head-quarters station is Mūsā Khel Bāzār; the only other place worth mention is Drug (population, 586). Fifty-four other villages are shown on the revenue rolls, but they seldom contain any permanent houses. Cultivation is in its infancy, and cattle-grazing is the chief occupation, the pasture grounds around Khajūri affording much fodder.

Bārkhān.—A *tahsīl* in the south-east of the Loralai District, Baluchistān, lying between 29° 37′ and 30° 21′ N. and 69° 3′ and 70° 4′ E., and bordering the Punjab, with an area of 1,317 square miles. Its population in 1901 was 14,922, an increase of 4,276 on the rough estimate made in 1891. The head-quarters station, which bears the same name as the *tahsīl*, lies about 3,650 feet above sea-level. The number of villages is 114. The land revenue in 1903–4 amounted to Rs. 47,000. The frequent existence of occupancy rights is a special feature of the tenures of the *tahsīl*. In the Leghāri-Bārkhān circle, one-third of the revenue levied is paid to the Leghāri chief as inferior proprietor of the soil, and he holds a revenue free-grant up to 1907. Bārkhān rugs are well-known, but have recently deteriorated in quality.

Duki Tahsīl.—A *tahsīl* of the Loralai District, Baluchistān, lying between 29° 53′ and 30° 25′ N. and 68° 12′ and 69° 44′ E., with an area of 1,951 square miles and population of 12,365 (1901), an increase of 4,356 since 1891. It lies from 3,000 to 5,000 feet above sea-level. The land revenue, including grazing tax, amounted in 1903–4 to Rs. 56,000. The head-quarters station, Duki, lies close to the village of that name. Villages number sixty-six. Some of the finest pasture grounds in Eastern Baluchistān are to be found here, and are visited by many Ghilzai Powindahs during the winter months.

Sanjāwi.—A sub-*tahsīl* of the Loralai District, Baluchistān, lying between 30° 9′ and 30° 28′ N. and 67° 49′ and 68° 35′ E., with an area of 446 square miles and population of 6,866 (1901), an increase of 1,334 since 1891. The head-quarters station, which bears the same name as the *tahsīl*, consists of a military fort occupied by the revenue establishment and local levies. Villages number thirty-seven. The land revenue, which is fixed in the case of irrigated lands, amounted in 1903-4 to Rs. 16,000. The Pechi Saiyids, who own lands in Pui, are exempted from payment of land revenue on certain conditions. Much of the *tahsīl* lies at an elevation of 6,000 feet above sea-level. Its glens, orchards, and gardens are very picturesque, and at Smāllan fine myrtle groves of great age are to be seen.

Bori.—A subdivision and *tahsīl* of the Loralai District, Baluchistān, lying between 30° 18′ and 30° 48′ N. and 67° 42′ and 69° 45′ E., with an area of 2,072 square miles and population of 18,174 (1901), an increase of 6,396 since 1891. The head-quarters station is at Loralai town (population, 3,561). The villages number 128. The land revenue amounted in 1903-4 to Rs. 61,000. Bori consists of a long valley, forming the catchment area of two branches of the Anambār river. It has rich soil and is well cultivated, and fine orchards are to be seen in some of the villages. The majority of the people are agriculturists. Among the Sargara Kākars of Dirgi a curious custom exists of allotting a share of land to every married woman at periodical distributions.

Loralai Town (*Loralī*).—A cantonment and, since 1903, the head-quarters station of the Loralai District, Baluchistān, situated in 30° 22′ N. and 68° 37′ E. It lies in the Bori *tahsīl*, 4,700 feet above sea-level, at the junction of the Harnai-Fort Sandeman and Pishīn-Dera Ghāzi Khān roads, about 55 miles from Harnai railway station. The climate is moderate, but high winds frequently prevail. It was selected and occupied as a military station in 1886. The population (1901) numbers 3,561, including a regiment of native cavalry and one of native infantry. Conservancy is provided for by a bazar fund, the income of which in 1903-4 was Rs. 10,900 and the expenditure Rs. 13,800. In the same year the income of the cantonment fund, which receives one-third of the net revenue from the octroi levied for the bazar fund, amounted to Rs. 10,900. In 1901 a piped water-supply of 75,000 gallons per diem was provided, at a total cost of about $1\frac{1}{4}$ lakhs of rupees.

Quetta-Pishīn.—A highland District of Baluchistān, lying Boun-
between 29° 52′ and 31° 18′ N. and 66° 15′ and 67° 48′ E., daries,con-
with an area of 5,127 square miles. It is bounded on the north and hill
and west by Afghānistān ; on the east by the Zhob and Sibi and river
Districts ; and on the south by the Bolān Pass and the systems.
Mastung *niābat* of the Kalāt State. The District consists of
a series of valleys of considerable length but medium width,
forming the catchment area of the Pishīn Lora, and enclosed
on all sides by the mountains of the TOBA-KĀKAR and
CENTRAL BRĀHUI ranges. The valleys vary in elevation from
4,500 to 5,500 feet, and the mountains from about 8,000 to
11,500 feet. On the north lie the Toba hills, containing the
fine plateaux of Loe Toba and Tabīna. This range sends out
the Khwāja Amrān offshoot southward to form the western boundary of the District under the name of the Sarlath. On the east
a barrier is formed by the mass of Zarghūn (11,738 feet), with
the ranges of Takatu (11,375 feet) and Murdār (10,398 feet).
Directly to the south lie the Chiltan and Mashelakh hills.
Besides the PISHĪN LORA, which, with its tributaries, drains the
greater part of the District, the only river of importance is the
Kadanai on the north, which drains the Toba plateau and
eventually joins the Helmand in Afghānistān. The District is
subject to earthquakes. Severe shocks occurred in December,
1892, and in March, 1902.

Two different systems of hill ranges meet in the neighbour- Geology.
hood of Quetta, giving rise to a complicated geological structure.
The principal rock formations belong to the permo-carboni-
ferous ; upper trias ; lias ; middle jurassic (massive limestone) ;
neocomian (belemnite beds) ; upper cretaceous (Dunghān) ;
Deccan trap ; middle eocene (Khojak shales, Ghāzij, and
Spīntangi) ; oligocene (upper Nāri) ; middle and upper miocene
(lower, middle, and upper Siwāliks) ; and a vast accumulation
of sub-recent and recent formations.

Except parts of the Toba, Zarghūn, and Mashelakh ranges, Botany.
the hills are almost entirely bare of trees. In the valleys are
orchards of apricot, almond, peach, pear, pomegranate, and
apple trees, protected by belts of poplar, willow, and *sinjid*
(*Elaeagnus angustifolia*). The plane (*chinār*) gives grateful
shade in Quetta. In spring the hill-sides become covered for
a little while with irises, red and yellow tulips, and many
Astragali. In the underground water-channels maiden-hair
fern is found. The valley basins are covered with a scrub
jungle of *Artemisia* and *Haloxylon Griffithii*. In parts *Tamarix gallica* covers the ground, and salsolaceous plants are

frequent. The grasses are chiefly species of *Bromus, Poa*, and *Hordeum*. On the Khwāja Amrān range wild rhubarb (*Rheum Emodi*) is found in years of good rainfall.

Fauna.   The 'reserved' forests in Zarghūn form a welcome breeding-ground for mountain sheep and *mārkhor*, but elsewhere they are decreasing in numbers. The leopard is found occasionally. A few hares are met with in the valleys. Wolves sometimes cause damage to the flocks in winter, and foxes are fairly abundant. Ducks are plentiful in the irrigation tanks in Pishīn. *Chikor* and *sīsī* abound in years of good rainfall.

Climate, temperature, and rainfall.   The climate is dry; dust-storms are common in the spring and summer months, especially in that part of the Chaman subdivision which borders on the Registān or sandy desert. The seasons are well-marked, the spring commencing towards the end of March, the summer in June, the autumn in September, and the winter in December. Only in July and August is the day temperature high; the nights are always cool. The mean temperature in summer is 78° and in winter 40°. The higher elevations are covered with snow in winter, when piercing winds blowing off the hills reduce the temperature below freezing-point. The total annual rain and snowfall varies from less than 7 inches in Chaman to 10½ in Quetta. Most of it is received between December and March.

History and archaeology.   In former times Pishīn was known as Fushanj and Pashang. The ancient name of Quetta was Shāl, a term by which it is still known among the people of the country, and which Rawlinson traces back to the tenth century. The district was held in turns by the Ghaznivids, Ghorids, and Mongols, and towards the end of the fifteenth century was conferred by the ruler of Herāt on Shāh Beg Arghūn, who, however, had shortly to give way before the rising power of the Mughals. The *Ain-i-Akbarī* mentions both Shāl and Pishīn as supplying military service and revenue to Akbar. From the Mughals they passed with Kandahār to the Safavids. On the rise of the Ghilzai power in Kandahār at the beginning of the eighteenth century, simultaneously with that of the Brāhuis in Kalāt, Quetta and Pishīn became the battle-ground between Afghān and Brāhui, until Nādir Shāh handed Quetta over to the Brāhuis about 1740. The Durrānis and their successors continued to hold possession of Pishīn and Shorarūd till the final transfer of these places to the British in 1879. On the advance of the army of the Indus in 1839, Captain Bean was appointed the first Political Agent in Shāl, and the country was managed by him on behalf of Shāh Shujā-ul-mulk. In March, 1842,

General England was advancing on Kandahār with treasure for General Nott when he was worsted in an encounter at Haikalzai in Pishīn, but the disgrace was wiped out at the same place a month later. The country was evacuated in 1842 and handed over to Kalāt. After Sir Robert Sandeman's mission to Kalāt in 1876, the fort at Quetta was occupied by his escort and the country was managed on behalf of the Khān up to 1883, when it was leased to the British Government for an annual rent of Rs. 25,000. It was formed, with Pishīn and Shorarūd, into a single administrative charge in 1883. Up to 1888 Old Chaman was the most advanced post on the frontier, but, on the extension of the railroad across the Khwāja Amrān, the terminus was fixed at its present site, 7 miles from that place. The boundary with Afghānistān was finally demarcated in 1895–6.

Many mounds containing pottery are to be found throughout the District. In the Quetta *tahsīl* the most ancient *kārez* are known to the people of the country as *Gabri*, i. e. Zoroastrian. While the present arsenal at Quetta was being excavated in 1886, a bronze or copper statuette of Hercules was unearthed, which was 2¼ feet high and held in its left hand the skin of the Nemean lion.

The number of towns is three, the largest being QUETTA, and of villages 329. The population was 78,662 in 1891 and 114,087 in 1901, an increase of 45 per cent. The following table gives statistics of area, &c., by *tahsīls* in 1901 :— *The people their tribes and occupations.*

Tahsīl.	Area in square miles.	Number of		Population.	Population per square mile.
		Towns.	Villages.		
Chaman . . .	1,236	1	4	16,437	13
Pishīn . . .	2,717	1	271	51,753	19
Quetta . . .	540	1	47	44,835	83
Shorarūd . .	634	...	7	1,062	2
Total	5,127	3	329	114,087	22

More than 84 per cent. of the people are Muhammadans of the Sunni sect ; Hindus number 10 per cent.; and Christians, who are chiefly Europeans, about 3 per cent. The language most widely spoken is Pashtū ; Brāhui is the tongue of about 6 per cent. of the people, and a little Persian is also used. Of the indigenous population 67,600, or 78 per cent., are Afghāns, rather more than half of them being Kākars and a third Tarīns.

Of the latter, the most numerous are the Abdāls, represented by the Achakzais occupying the Chaman subdivision and part of Pishīn. The Brāhuis, who live in the south of the District, form 8 per cent., and Saiyids, who are numerous in the Pishīn *tahsīl*, about 9 per cent. The indigenous population is almost entirely engaged in cultivation and flock-owning. The Afghāns of Pishīn, especially the Huramzai Saiyids, carry on a large trade in horses. Many of them have made their way as far as Australia, or are engaged in trade in parts of India.

Christian missions.   The missions working in Quetta consist of branches of the Church Missionary Society and of the Church of England Zanāna Missionary Society. They maintain two hospitals and four schools, one of which is aided by Local funds. A mission church was opened in 1903. The efforts of the workers are principally devoted to medical aid and education, and few converts have so far been made among the people of the country.

General agricultural conditions.   The soil in the centre of the valleys consists of fine clay and sandy beds. Along the skirts of the hills loess is found, and higher up a fringe of coarse-grained gravel. The soil of Shorarūd is impregnated with salt. At Barshor, in the Pishīn *tahsīl*, cultivation is carried on in terraced fields. Crops are assured only on lands which can be permanently irrigated. The 'dry' crop area consists chiefly of embanked land to which flood-water is led. Irrigated land is allowed to lie fallow for one to three years, unless it can be manured; 'dry' crop land can be cultivated every year, but more than one good crop in five years is seldom obtained. The harvest reaped in spring is sown with the help of the winter rains; the autumn harvest, which is small compared with the former, is sown in June and July.

Agricultural statistics and principal crops.   The cultivable area in the two *tahsīls* of Quetta and Pishīn, which have been cadastrally surveyed, is 706 square miles, of which 324 are cultivated by rotation. Of this latter total, 221 square miles (68 per cent.) are permanently irrigated (*ābi*); and the remainder are either flood-crop (*sailāba*) or 'dry' crop (*khushkāba*). The area under crop in 1902–3 was 72 square miles, of which 79 per cent. was under wheat, the staple grain of the District; 4 per cent. under barley; 10 per cent. under maize and millets; 3 per cent. under green vegetables; and 4 per cent. under lucerne. Owing to the peace and protection which have followed the British occupation, cultivation has increased very largely during the past twenty-five years. Potatoes, vegetables, and lucerne are profitably cultivated;

fruit orchards and vineyards are extending ; and great attention is bestowed on melon growing. The cultivators eagerly avail themselves of Government loans, the amount advanced between 1897 and 1904 being 1·3 lakhs.

The short-legged breed of Kachhi cattle is imported for Cattle, the plough. Transport is by camel, and these animals are horses, used in the plough in Chaman and Pishīn. The local breed sheep, &c. of horses is excellent, and has been much improved by the introduction of imported stallions, of which eighteen are generally stationed in the District in summer. The branded mares number 256. A horse-fair and cattle-show is held at Quetta in the autumn, which is largely patronized by local breeders. Sheep imported from Siāhband in Afghānistān are much prized.

Of the total irrigated area in the *tahsīls* of Quetta and Irrigation. Pishīn, 14 per cent. is supplied from Government irrigation works and 66 per cent. from the 254 *kārez* or underground channels. Water is also obtained from 18 streams and 854 springs. Artesian wells number twenty-four. The Government irrigation works are the Khushdil Khān reservoir and the Shebo canal, both situated in Pishīn. The former, which is fed by flood-water from two feeder-cuts, is capable of holding about 750 million cubic feet of water. It commands about 17,000 acres, but the average area cultivated by its aid has hitherto been only 3,300 acres. The area will probably be increased by improvements effected in 1902. Up to 1903 the capital cost incurred was about 10 lakhs. The Shebo canal takes off from the Quetta Lora and is supplemented by a system of tanks. It commands 5,340 acres, but less than half of this is irrigated annually. The capital cost up to 1903 was about 6¾ lakhs. Revenue and water-rate are levied together, on both systems, in the shape of one-third of the gross produce, the whole amount being credited to the Irrigation department.

In 1903 the District contained four juniper Reserves on the Forests. Zarghūn range, with an area of 52 square miles ; two pistachio forests of 13 square miles ; and one mixed forest covering 2 square miles. In the latter tamarisk is the chief tree. Experimental plantations, covering 63 acres, are maintained close to Quetta.

Coal is found in the Sor range to the east of Quetta. The Minerals. seam is narrow, but has been traced for nearly 20 miles. It is worked in different places by five contractors. The output, which is entirely consumed in Quetta, was 7,148 tons in

1903. Chromite has been discovered in scattered pockets in the serpentines and basic igneous intrusions near Khānozai, for working some of which a lease has been given to the Baluchistān Mining Syndicate. During 1903 about 284 tons were extracted.

Arts and manufactures. The manufacture of felts and of rugs formed by the *darī* stitch is an indigenous industry. Excellent silk embroidery is prepared, especially by Brāhui women. In Quetta, Kandahāris make copper vessels, which are equal in quality to those sold in Peshāwar. The Murree Brewery Company has a branch at Kirāni, about 5 miles from Quetta, the output of which was 347,220 gallons of beer in 1903. In 1904 some successful experiments were made in sericulture.

Commerce. The great increase in trade is referred to in the article on QUETTA TOWN. The only other marts of importance are Kila Abdullah and Chaman, from both of which places trade is carried on with Afghānistān. The total value of this trade amounted to about 13½ lakhs in 1903, imports being valued at 6½ and exports at 7 lakhs. Live animals, *ghī*, asafoetida, fresh and dried fruit, and pile carpets are the principal imports from Afghānistān, and food-grains, piece-goods, and metals from India. Exports to India are chiefly wool, *ghī*, and fruit, and to Afghānistān piece-goods, metals, and dyes.

Railways and roads. The Mushkāf-Bolān branch of the North-Western Railway, on the standard gauge, enters the District from the south and runs to Quetta, where it meets a branch of the Sind-Pishīn section from Bostān. The latter line enters the District near Fuller's Camp and runs across the Pishīn plain to Chaman. The District is well provided with roads, the total length of metalled and partially metalled roads being 405, and of unmetalled paths 228 miles. They are maintained partly from Provincial revenues and partly from Military funds.

Famine. Owing to its large irrigated area and excellent communications, the District is well protected and actual famine has not been known. Some distress occurred between 1897 and 1902, owing to deficient rainfall and to damage done by locusts. Relief was afforded by the suspension and remission of land revenue, the grant of advances for the purchase of seed-grain and bullocks, and the opening of relief works, costing about Rs. 14,000. In years of deficient pasturage the railway is used by graziers to transport their flocks to more favoured tracts.

District subdivisions and staff. The District is divided into three subdivisions and *tahsīls*: CHAMAN, PISHĪN, and QUETTA. Of these, Chaman, Pishīn, and Shorarūd in Quetta form part of British Baluchistān, and the rest of the Quetta *tahsīl* is Agency Territory. The executive head of the District combines the functions of Deputy-Commissioner for areas included in British Baluchistān, and of Political Agent for Agency Territories. A Native Assistant is in charge of Chaman, an Extra Assistant Commissioner of Pishīn, and the Assistant Political Agent of the Quetta subdivision. The *tahsīls* of Quetta and Pishīn each have a *tahsīldār* and a *naib-tahsīldār* for revenue work. The superior staff at head-quarters includes a Superintendent of Police, two Extra Assistant Commissioners, a Cantonment Magistrate, and an Assistant Cantonment Magistrate.

Civil work at Quetta is disposed of by a Munsif, and four Honorary Magistrates assist the ordinary staff in deciding criminal cases. Both civil and criminal powers are exercised by all the officers mentioned in the preceding paragraph. The Political Agent is the District and Sessions Judge. In 1903 the total number of cognizable cases reported was 1,402, conviction being obtained in 1,232. Most of the cases were of a petty nature. The total number of criminal cases disposed of by the courts in 1903-4 was 3,102, and of civil cases 4,807. Disputes were referred to a *jirga* for award under the Frontier Crimes Regulation in 203 cases. *Civil justice and crime.*

The District furnished the emperor Akbar with a force of 2,550 horse and 2,600 foot; Rs. 750 in cash; 4,340 sheep; 1,280 *kharwārs* of grain, and 7 maunds of butter. Nādir Shāh assessed Pishīn to furnish a fixed number of men-at-arms, a system known as *gham-i-naukar*, which was continued by Ahmad Shāh Durrani, in whose time 895 *naukars* were taken. In the time of Tīmūr Shāh some of the tribesmen were recalcitrant, and the land of 151 *naukars* was confiscated. The remaining service grants were subsequently commuted for cash payment. When the District came into the hands of the British this cash payment was still in force in some parts of the Pishīn *tahsīl*, while in others the system had broken down, and *batai*, or the taking of an actual share of the produce, had been substituted. The combined system was continued in Pishīn up to 1889, the Government share of the produce being levied at rates varying from one-third to one-sixth. In 1899 a fixed cash assessment on irrigated estates was introduced for twenty years. The incidence per irrigated acre ranged from a maximum of Rs. 5-0-3 to a minimum of Rs. 1-5-3, the average being Rs. 2-13-10. In the Quetta valley, the land revenue under native rule was obtained partly from a fixed assessment in cash or kind, called *zar-i-kalang*, partly *Land revenue administration.*

from appraisement, and partly by division of the crops. The system continued up to 1890, when *batai* at a uniform rate of one-sixth of the produce and grazing tax were introduced. A fixed cash assessment was imposed on irrigated lands for ten years from 1897, and is now about to be revised. The maximum incidence per acre on irrigated area was Rs. 3–9–4, the minimum Rs. 1–6–2, and the average Rs. 2–0–4. In Shorarūd, revenue was first levied in 1882–3 at one-sixth of the produce, and from April, 1897, a fixed cash assessment was imposed on irrigated lands. Large revenue-free grants are held, especially in Pishīn. The estimated annual value of the land revenue thus alienated is Rs. 42,700. The total land revenue of the District in 1903–4 was 1·5 lakhs, and the revenue from all sources 3·2 lakhs. The land revenue yielded 47 per cent. of this total, stamps 12 per cent., and excise 35 per cent.

Local boards.
    The Quetta municipality was formally constituted in October, 1896. Its affairs are managed by a committee, consisting of thirteen nominated official and non-official members, with the Political Agent as *ex officio* president. The only Local fund is the Pishīn Sadar and District bazar fund, which is controlled by the Political Agent. Its chief source of income is octroi, and its expenditure is incurred on objects of public utility, principally at Pishīn and Chaman. The income in 1903–4 amounted to Rs. 39,600 and the expenditure to Rs. 34,000.

Army.
    QUETTA is the head-quarters of the fourth division of the Western Command and has the usual staff. Besides the garrison of Quetta, a native infantry regiment is stationed at Chaman and detachments are posted at Pishīn and, to guard the Khojak tunnel, at Shelabāgh and Spinwāna.

Police and jails.
    In 1904 the total force of police amounted to 519 men, of whom 362 were constables and 53 horsemen. The officers include a District Superintendent, an Assistant Superintendent, five inspectors, and eleven deputy inspectors. The force was distributed in seventeen stations. The Quetta municipality pays for a force of eighty-six police, the cantonment committee for eighty-four, and Local funds for twenty-four watchmen. The local levies number 487, including 170 mounted men. There is a District jail at Quetta, and a subsidiary jail at Pishīn, with total accommodation for 139 male and 10 female prisoners. Convicts whose term exceeds six months are generally sent to the Shikārpur jail in Sind.

Education.
    In educational, as in other respects, the District is the most

advanced in the Province. In 1904 the number of Government and aided schools was twelve, with 827 pupils, including 148 Indian girls and 44 European and Eurasian children. The cost amounted to Rs. 23,500, of which Rs. 7,700 was derived from fees and subscriptions, and Rs. 7,100 from Provincial revenues, the balance being met by the North-Western Railway and from Local funds. The three mission schools had eighty-five pupils. About 900 pupils were under instruction in mosque schools.

The District possesses one Government-aided hospital, in charge of a Civil Surgeon, and seven dispensaries, including a female dispensary maintained from the Lady Dufferin fund. They contain accommodation for 118 in-patients. The total attendance of patients in 1903 was 63,310; the daily average attendance in Government institutions of in-patients being 59 and of out-patients 211. Two of these institutions are maintained by the North-Western Railway, at Bostān and Shelabāgh, and two receive grants from Local funds; the expenditure of the others is met from Provincial revenues. In 1903 the total expenditure from Provincial revenues and Local funds amounted to Rs. 18,109. The Church of England Medical Mission maintains two hospitals, to which 592 in-patients were admitted in 1902, while the out-patients numbered 19,190. *(Hospitals and dispensaries.)*

Vaccination is compulsory in the town and cantonment of Quetta, and there are indications that the people are beginning to prefer this method to inoculation. The number of successful vaccinations in 1903 was 2,660, or about 23 per thousand of the population. *(Vaccination.)*

[*Settlement Report of the Pishīn Tahsīl* (1899).—J. H. Stocqueler: *Memorials of Afghānistān.* (Calcutta, 1843.)—*Records of the Geological Survey of India*, vol. xxvi, pt. 2 of 1893.]

Chaman Subdivision.—The most northerly subdivision and *tahsīl* of the Quetta-Pishīn District, Baluchistān, lying between 30° 28′ and 31° 18′ N. and 66° 16′ and 67° 19′ E. It is bordered on the north by Afghānistān. The greater part consists of the mountainous region called Toba, which has a mean elevation of about 8,000 feet, though its western skirts descend to about half that height. There is little cultivation, pasture being the principal means of livelihood. The area is 1,236 square miles, and the population in 1901 was 16,437, showing an increase of 5,375 since 1891. The only place of importance is the head-quarters, CHAMAN TOWN, population (1901) 2,233. The indigenous Achakzai Afghāns are nomadic, and permanent villages are practically unknown. They pay as revenue a lump assessment of Rs. 8,000 per annum.

Pishīn.—A subdivision and *tahsīl* covering the centre of the Quetta-Pishīn District, Baluchistān, lying between 30° 1' and 31° 12' N. and 66° 21' and 67° 48' E. It consists of the southern slopes of the Toba hills and the basin of the Pishīn Lora, the latter being a plain lying about 5,000 feet above sea-level. The area of the *tahsīl* is 2,717 square miles; its population in 1901 was 51,753, showing an increase of 14,573 since 1891. Pishīn, the head-quarters, which has sprung up since the British occupation, is 6 miles from Yāru Kārez railway station. The villages number 271, and the land revenue in 1903-4 amounted to Rs. 80,700. Large revenue-free grants, a relic of Afghān rule, are held chiefly by Saiyids. The *tahsīl* contains two irrigation works, the Shebo canal and the Khushdil Khān reservoir.

Quetta Subdivision.—A subdivision and *tahsīl* of the Quetta-Pishīn District, Baluchistān, lying between 29° 52' and 30° 27' N. and 66° 15' and 67° 18' E. It is held on a perpetual lease from the Khān of Kalāt. For administrative purposes Shorarūd, which is British territory, is attached to it. The two cover an area of 1,174 square miles, of which 540 form the Quetta *tahsīl* proper. The population in 1901 numbered 45,897, that of Shorarūd being 1,062. The only town is QUETTA (population, 24,584); and the villages number fifty-four. The *tahsīl* occupies a valley about 5,500 feet above sea-level, surrounded by mountains. Shorarūd derives its name from a stream of brackish water, which traverses it to join the Pishīn Lora; it consists of the river basin and the Sarlath hills, beyond which lies Shorāwak in Afghānistān. The Sarlath hills afford excellent pasturage. Shorarūd contains only seven permanent villages. The land revenue of the whole *tahsīl* in 1903-4 amounted to Rs. 65,500, of which Rs. 2,000 was contributed by Shorarūd. Owing to the ready market available in the Quetta town and cantonment and the numerous *kārez*, the Quetta valley is the best cultivated in Baluchistān, and the extension of fruit gardens has been marked. Coal is found in the adjoining Sor range. A branch of the Murree Brewery has been worked near Kirāni since 1886.

Chaman Town.—Head-quarters of the Chaman subdivision of the Quetta-Pishīn District, Baluchistān, and the frontier terminus of the North-Western Railway, situated in 30° 56' N. and 66° 26' E., at an elevation of 4,311 feet above the sea. It is the head-quarters of a Native Assistant. Population (1901), 2,233. The garrison consists of a regiment of native infantry and some cavalry occupying the fort. A supply of water is brought in pipes from the Bogra stream, the system having cost 2¼ lakhs. The conservancy of the civil station is provided for from the Pishīn bazar fund.

Khojak (*Kozhak*).—An historic pass across the Khwāja Amrān offshoot of the TOBA-KĀKAR mountains in the Quetta-Pishīn District, Baluchistān. It lies in 30° 51′ N. and 66° 34′ E., 70 miles from Quetta by rail. From Kila Abdullah, on the south, there is a gradual ascent to Shelabāgh, whence the summit (7,457 feet) is reached in 3¾ miles. A cart-road through the pass connects Kila Abdullah with Chaman. At Shelabāgh the railway runs through the Khojak tunnel, which is just under 2½ miles long, and cost rather less than 70 lakhs of rupees, or about Rs. 530 per lineal foot. It was constructed between 1888 and 1891. Lying on the route from Kandahār to India, the Khojak Pass has been crossed and recrossed for centuries by conqueror, soldier, and merchant; and its passage was twice effected by the British arms, in 1839 and in 1879.

Quetta Town (*Kwatah*, locally known as Shāl or Shālkot). —Capital of the Baluchistān Agency and head-quarters of the Quetta-Pishīn District, situated in 30° 10′ N. and 67° 1′ E., at the northern end of the *tahsīl* of the same name. It is now one of the most desirable stations in Northern India. Quetta is connected with India by the North-Western Railway, being 727 miles from Lahore and 536 from Karāchi. It was occupied by the British during the first Afghān War from 1839 to 1842. In 1840 àn assault was made on it by the Kākars, and it was unsuccessfully invested by the Brāhuis. The present occupation dates from 1876. The place consists of the cantonment on the north, covering about fifteen square miles, and the civil town on the south, separated by the Habīb Nullah. Population has risen from 18,802 (1891) to 24,584 (1901). It includes 3,678 Christians, mainly the European garrison, 10,399 Muhammadans, and 8,678 Hindus. The majority of the remainder are Sikhs. The ordinary garrison comprises three mountain batteries, two companies of garrison artillery, two British and three native infantry regiments, one regiment of native cavalry, one company of sappers and miners, and two companies of volunteers. The police force employed in the cantonment and town numbers 180.

Municipal taxes have been levied since 1878, but the present municipal system dates from 1896. The income in 1903–4 was 2·2 lakhs, chiefly derived from octroi, and the expenditure was 2·1 lakhs. The committee has obtained loans from Government for carrying out drainage and water-works, of which the unpaid balance on March 31, 1904, amounted to Rs. 31,100. Half of the net octroi receipts is paid over to the cantonment fund. The receipts of this fund, from which the maintenance of the cantonment is provided, were 1·1 lakhs in 1903–4 and the expenditure 1·3 lakhs. Much attention has been paid to sanitation and the prevention of enteric fever, which was at one time common. A piped supply of water for the cantonment, civil station, and railway was completed in 1891 at a cost of about $7\frac{1}{4}$ lakhs, and an additional supply has since been provided for the cantonment at a cost of more than $3\frac{1}{4}$ lakhs. The civil station and town lie somewhat low, and nearly $1\frac{1}{4}$ lakhs has been expended in providing a system of street drainage. The principal buildings are the Residency, the Sandeman Memorial Hall, St. Mary's Church, and the Roman Catholic Church. The civil hospital is well equipped, and the town also possesses a female dispensary, two mission hospitals, a high school, a girls' school, and a European school. A mill for grinding flour and pressing wool and chopped straw has existed since 1887. The Indian Staff College is now in course of erection (1906). A feature of the station is the gymkhana ground, with its fine turfed polo and cricket grounds. The trade of Quetta is continually expanding. Imports by rail have increased from 39,200 tons in 1893 to 56,224 tons in 1903, and exports from 5,120 to 13,829 tons.

Boundaries, configuration, and hill and river systems.

Chāgai District.—A District in Baluchistān, lying between 28° 2′ and 29° 54′ N. and 60° 57′ and 66° 25′ E., with an area of 18,892 square miles. It is bounded on the north by Afghānistān; on the east by the Sarawān division of the Kalāt State; on the south by Khārān; and on the west by Persia. From Nushki westward to Dālbandin the country consists of a level plain, having a slight westerly slope, with sand-dunes in the centre and on the north. Beyond Dālbandin lies the western corner of the Baluchistān desert, consisting generally of pebbly plains on the south and sandy desert on the north. Most of the sand-hills of this desert are shaped alike, being in the form of a crescent with the horns to the south and the curve to the north. They vary a good deal in height, the tops of the largest being about sixty feet above the plain and sloping down gradually to the horns, where they mingle with the sand. A single line of such hills stretches across the gravel plain from

Mashki Chāh to Reg-i-Wakāb. North of Dālbandin lie the
CHĀGAI hills, and to the south is the Rās Koh. The largest
river is the PISHĪN LORA, known locally as the Dhor. The
Kaisār stream debouches into the plain near Nushki. Many
hill-torrents carry off the drainage of the Chāgai hills, the
chief being the Morjen. The District contains the swamp
known as the HĀMŪN-I-LORA.

The geology displays characteristic desert formations, such Geology.
as the dried-up beds of salt lakes surrounded by successive
tiers of shingle terraces; level flats of dried mud called *pat*;
plains strewn with pebbles called *dasht*; the gigantic talus or
dāman which half buries the straggling hill-ranges; and finally
the gradual accumulation of wind-borne sand. The hill ranges
contain an interesting series of rocks, in which many of the
geological strata characteristic of Baluchistān have been
recognized. Volcanic rocks of the Deccan trap period are
well displayed.

The vegetation consists, for the most part, of a scanty and Botany.
ill-favoured scrub. Pistachio and tamarisk are the only trees.
The saline soil produces varieties of *Salsola* and *Haloxylon*;
and in the sand-dunes, among other plants, two varieties of
tamarisk and *Euphorbia* occur. The commonest of the occasional grasses are *Eragrostis cynosuroides, Aristida plumosa,*
and several species of *Aeluropus*. Asafoetida occurs on the
Koh-i-Sultān.

The deserts swarm with venomous snakes and scorpions, Fauna.
while skinks (*reg māhi*) are found in the sand-hills. In the
remotest parts the wild ass occurs, and the Persian gazelle is
fairly common.

The climate is dry and agreeable in the autumn and spring. Climate,
From May to September great heat is experienced by day, but tempera-
the nights are cool. The western half of the District is at this rainfall.
time exposed to the effects of the *bād-i-sad-o-bīst-roz*, or 120
days' wind, which carries with it clouds of sand. The winter
is cold. Much sickness is caused by the presence of sulphates
in the water, which is often fit for consumption only after distil-
lation. Between 2 and 3 inches of rain are received, chiefly
in winter. Snow falls on the hills.

Local tradition speaks of an Arab and Mongol occupation History
of the country in early times. In 1740 Nādir Shāh conferred and
Nushki as a fief on the chief of Khārān, but it fell into the logy.
hands of the Brāhuis shortly afterwards and became a *niābat*
of the Kalāt State. Henry Pottinger visited the country in
1810, and Sir Charles Macgregor in 1877. In 1886 the Amir

of Afghānistān sent a force to occupy Chāgai; but ten years later it fell, with Western Sinjrāni, within the British sphere, by the decision of the Afghān-Baloch Boundary Commission, and an Assistant Political Agent was thereon placed in charge of the country from Nushki to Robāt Kila. In 1897 the transit dues levied by the Zagar Mengal chief were abolished, in consideration of an annual payment of Rs. 7,000, of which Rs. 3,600 is devoted to a Mengal levy service. Finally, in June, 1899, the Nushki *niābat* was leased to the British Government by the Kalāt State for an annual quit-rent of Rs. 9,000, and a *tahsīl* was established. In 1901 a sub-*tahsīl* was located at Dālbandin in Chāgai.

The people, their tribes and occupations.
The population enumerated in 1901 was 15,689; allowing 6,000 for Western Sinjrāni, where no census was taken, the total is 21,689. The following table gives statistics of area, &c., by *tahsīls* in 1901 :—

| *Tahsīl.* | Area in square miles. | Number of | | Population. | Population per square mile. |
		Towns.	Villages.		
Nushki . . .	2,202	...	10	10,756	5
Chāgai sub-*tahsīl* .	7,283	...	22	4,933	One person to 1·5 square miles.
Western Sinjrāni .	9,407	...	...	Not enumerated.	
Total	18,892	...	32	15,689	1

Nearly all the inhabitants are Muhammadans of the Sunni sect. The languages spoken are Brāhui, Baluchi, and a little Pashtū. The majority of the people are Brāhuis, the principal Brāhui tribes being Zagar Mengals (4,600) and Muhammad Hasnis (4,300). The Rakhshānis, another important tribe (3,500), claim to be Baloch. The tribes living in Chāgai and the Western Sinjrāni country include Sinjrānis, Dāmanis, Kūrds, and Rekis. In Nushki most of the population are cultivators; elsewhere they are chiefly flock-owners. The permanent villages number thirty-two, including NUSHKI, the head-quarters station.

Agriculture and irrigation.
Agriculture is in its infancy, and the people are wanting in industry. The cultivated area is considerable only in the Nushki *tahsīl*, but it depends chiefly on uncertain floods. In the *dāk* or alluvial plain the clay soil is very fertile when irrigated. Between 1899–1900 and 1903–4, about Rs. 28,500 was advanced

for the encouragement of cultivation. Large numbers of camels, sheep, and goats are bred. Bullocks and ponies are scarce. The only sources of permanent irrigation are *kárez* and streams, the former numbering twenty-one and the latter seven. In rainy years water raised by means of dams from the Pishīn Lora is capable of irrigating immense areas in the *dāk* lands in Nushki; and, with this object, a dam in the Bur Nullah was constructed by Government in 1903, at a cost of Rs. 14,000, but it was washed away in 1904. It has since been replaced.

A small establishment is maintained for the protection of Forests and the pistachio and tamarisk jungles in the hills east of Nushki. minerals. Lead, copper and iron ores, sulphur, gypsum, alunogen, and various ornamental stones, such as Oriental alabaster, occur abundantly, but, owing to the inaccessibility of the region, are of no present industrial value.

The women make a few rugs in the *darī* stitch for home Manufac-use, which are of good texture and pattern and have fast tures and dyes. Trade converges on Nushki from Seistān, Afghānistān, commerce. and Khārān ; and the District has acquired much of its importance from the opening of the Nushki-Seistān trade-route connecting India with Persia. During the five years ending in 1904 the average annual value of the total trade on this route was 14.1 lakhs, the exports averaging 7.2 lakhs and the imports 6.9 lakhs. The District itself produces a little wool, *ghī*, and asafoetida.

A branch of the North-Western Railway on the standard Railways gauge was opened from Spezand to Nushki in 1905. The and roads. Nushki-Seistān trade-route has nineteen stages in the 327 miles between Nushki and Robāt Kila, the frontier station. Shelters have been provided for travellers throughout, and post and telegraph offices and shops are located at important stages. The distance from Robāt Kila to Nasratābād, the capital of Seistān, is 106 miles.

The peculiar conditions of the country render it liable to Famine. prolonged and constant droughts, at which times the people migrate to other parts. The central and western tracts at present depend entirely on the Helmand valley for their grain supply. In 1902 advances amounting to Rs. 4,200 were given for the purchase of seed and plough bullocks.

The District forms part of the Agency Territories, and is District administered under the executive orders of the Governor-sub-General. No laws have yet been formally applied. It is staff, &c. divided into three parts : the NUSHKI *tahsīl*, the CHĀGAI sub-*tahsīl*, and the Western Sinjrāni country. The executive authority is vested in a Political Assistant, who is assisted by a European Assistant District Superintendent of Police, one *tahsīldār*, and two *naib-tahsīldārs*. The last three officials are magistrates, with jurisdiction in petty civil suits. Most cases are, however, referred to *jirgas* or councils of elders. Such cases numbered twenty-five in 1903–4, including one murder and one robbery. The number of criminal cases in the same year was four and of civil suits six.

Land revenue administration.
In Nushki the Kalāt system of levying land revenue at the rate of one-tenth of the produce together with certain cesses has been continued, with certain modifications. Government owns shares in the water of most of the *kārez*. In Chāgai the rate of revenue levied is one-sixth of the produce. Grazing tax is levied in Chāgai and also on trans-frontier animals grazing in the Nushki *tahsīl*. A tax in kind is collected on asafoetida. In 1903–4 the land revenue of the District amounted to Rs. 17,000, and the revenue from all sources to Rs. 26,000.

Army, police, jails, education, and medical.
A detachment of infantry is stationed at Nushki. The local levies, under the Assistant District Superintendent of police, numbered 186 in 1904. The regular police force comprises only eleven men, and has been amalgamated since 1903 with the Quetta-Pishīn police. Four watchmen are paid from Local funds. The lock-up at Nushki has accommodation for twenty male and four female prisoners. A primary school, founded at Nushki in 1904, is attended by twenty boys, and sixty boys receive instruction in the village mosques. A civil dispensary was opened in 1900, with accommodation for eight in-patients. The average daily attendance of all patients in 1903 was thirty-three. The expenditure, which was borne by Provincial revenues, amounted to Rs. 1,648. Vaccination has not been attempted, and the people still resort to inoculation.

['A Geological Sketch of the Baluchistān Desert,' *Memoirs of the Geological Survey of India*, vol. xxxi, part 2.—*The Botany of the Afghān Delimitation Commission*, vol. iii (1887).—Sir C. Macgregor: *Wanderings in Baluchistān* (1882).]

Nushki Tahsīl.—The most easterly *tahsīl* of the Chāgai District, Baluchistān, lying between 29° 2' and 29° 54' N. and 65° 13' and 66° 25' E., at an elevation of about 3,000 feet. The area is 2,202 square miles, and the population (1901) 10,756. The eastern portion is hilly; the remainder consists of a level plain, with sand-hills on the north and in the centre. The great stretch of alluvial plain, known as the *dāk*, is very fertile when irrigated, but the absence of cultivators and of drinking-water presents much difficulty. The number of permanent villages is ten. Most of them are situated under the hills on the east. NUSHKI TOWN (population, 644) is the head-quarters station, from which the *tahsīl* gets its name. The revenue in 1903–4 amounted to Rs. 10,700.

Chāgai Sub-tahsīl.—A sub-*tahsīl* of the Chāgai District, Baluchistān, lying between 28° 19′ and 29° 34′ N. and 63° 15′ and 65° 35′ E., and bounded by Afghānistān on the north and the Rās Koh hills on the south. The area is 7,283 square miles, and the population (1901) 4,933. For revenue purposes it includes part of Western Sinjrāni, which covers 9,407 square miles and has an estimated nomadic population of 6,000. This tract consists almost entirely of pebbly plains and sandhills, and is probably the most uninviting area in the whole Province. The water, especially in summer, is impregnated with sulphates. The people of Chāgai are essentially pastoral, and so far have exhibited little aptitude for agriculture. They own large flocks of sheep and herds of camels. The permanent villages number twenty-two. Dālbandin is the head-quarters. In 1903–4 the land revenue amounted to Rs. 6,100, about half of which consisted of the proceeds of the grazing tax.

Nushki Town.—Head-quarters of the Chāgai District, Baluchistān, and terminus of the Seistān trade route, situated in 29° 34′ N. and 66° 0′ E., at an elevation of 2,900 feet above the sea, 91 miles from Quetta, with which it was connected by railway in 1905. Population (1901), 644. It came into existence in 1899, when the Nushki *niābat* was taken over from the Khān of Kalāt; and a bazar of some fifty shops and buildings for the *tahsīl*, police, and levy establishments quickly sprang up. The place is, however, inconveniently situated, as the water-supply from the Kaisār stream is liable to dry up almost completely in summer. A house and shop tax is levied, and the proceeds are used in maintaining sanitary establishments. The income of this fund in 1903–4 was Rs. 1,800 and the expenditure Rs. 1,300. Trade from Nushki is carried on with Khārān, with Garmsel and Shorāwak in Afghānistān, and with Jālk and Seistān in Persia.

Bolān Pass.—A District of Baluchistān, named after the Bcundaries, conhistoric pass, lying between 29° 24′ and 30° 10′ N. and figuration, 67° 4′ and 67° 44′ E., with an area of 896 square miles. and hill

and river
systems.

The pass proper extends from Kolpur, known to the natives as Kharlakai Kotal, to Rindli, and is about 54 miles long. It is widest in the Laleji plain on the south, whence it narrows to a gorge known as Afghān Ponzak. The elevation rises from 750 feet to about 5,900 feet. The District is bounded on the east by the Sibi District, and on its remaining sides by the Sarawān and Kachhi divisions of the Kalāt State, and is enclosed between high mountains belonging to the CENTRAL BRĀHUI range. The Bolān river rises near Kolpur, but the water makes its first appearance at Sar-i-Bolān and disappears again near Abigum. At Bībī Nāni it is joined from the west by the Sarawān river, and from this point possesses a perennial stream. Many hill-torrents empty themselves into the river, causing violent floods after heavy rain.

Geology.

The rocks consist of a varied series, including jurassic and lower and upper cretaceous strata; basalt flows of Deccan trap age; Ghāzij and Spīntangi beds (middle eocene); lower Nāri (upper eocene); lower, middle, and upper Siwāliks (middle and upper miocene); and recent and sub-recent deposits.

Botany.

The vegetation consists of a repellent scrub, made up of such plants as *Capparis aphylla, Acanthodium spicatum, Prosopis spicigera, Withania coagulans, Calotropis procera, Alhagi camelorum*, and three kinds of *Acacia*. On the surrounding hills occur pistachio and a little olive.

Fauna.

Sīsī and *chikor* are found in the upper parts of the pass, and a few hares and ravine-deer occur in the Laleji plain. Fish exceeding 20 lb. in weight have been caught with the rod in the lower reaches of the Bolān river.

Climate,
temperature, and
rainfall.

The climate varies with the elevation. In summer the heat in the lower parts is trying, while in winter snow falls above Mach. The average annual rainfall is about 8 inches. Most of it is received in winter, but an occasional fall occurs in July.

History
and
archaeology.

The Bolān Pass has for centuries been the route which traders, invaders, and nomad hordes have traversed between India and High Asia, and has been the scene of many battles between the people of the highlands and of the plains. In the early days of the British connexion with the country it was nominally under the control of the Khān of Kalāt; but the Kūrd and the Raisāni tribes had acquired rights to levy transit-dues, and it was a favourite raiding ground of the Marris and Kākars. The army of the Indus negotiated the pass without much opposition in 1839, and it was again traversed by the army for Southern Afghānistān in 1878.

In pursuance of the policy of freedom of trade between Kalāt and India, posts were established in the pass soon after the British occupation in 1877; and in 1883 the Khān of Kalāt ceded civil and criminal jurisdiction in the pass and his rights to levy tolls, in return for an annual payment of Rs. 30,000. The tolls were abolished in 1884, and allowances were given to the Raisāni, Kūrd, and other tribesmen who had shared in the proceeds of the transit-dues. The Bolān was first attached to the old Thal-Chotiāli District; it was then placed under Quetta-Pishīn; and finally, in 1888, under the Political Agent in Kalāt.

The District possesses only two permanent villages of any size, Mach and Kirta. The total population (1901) amounted to 1,936. The Kuchiks, a section of the Rind Baloch numbering 326, are the cultivating proprietors of the soil. The total area of cultivable land is 3,300 acres, about one-third of which is generally cropped each year. Most of the cultivation is at Kirta, which is irrigated from the permanent stream of the Bolān. The water and land are divided for each crop according to the number of adult males among the Kuchiks. The principal crop is wheat; some *jowār* also is cultivated in the summer. *The people and agriculture.*

Thin seams of coal in the Ghāzij strata near Old Mach are worked by a private firm. The output in 1903 amounted to 3,259 tons. In the spring of 1889 a boring for petroleum was put down near Kirta, and a show of oil was struck at 360 feet, but the boring was abandoned owing to an influx of hot sulphurous water. Good sulphur has also been discovered. No trade of importance exists. The Mushkāf-Bolān branch of the North-Western Railway enters the District at Nāri Bank station, and a road traverses the pass connecting Sibi with Quetta, which is metalled and bridged between Rindli and Quetta. *Minerals, railways, and roads.*

The District, which is officially known as the Bolān Pass and Nushki Railway District, forms part of the Agency Territories. Besides the pass and the civil station of Rindli, it includes jurisdiction over the road and railway from the Nāri river to a point within about 13 miles of Quetta, and over the portion of the Nushki Railway lying in Kalāt. The Political Agent, Kalāt, holds executive charge and has the powers of a District and Sessions Judge. The Assistant Political Agent, Kalāt, and the Native Assistant for the Sarawān country also have jurisdiction. The official in immediate charge of the pass is a *tahsīldār*, posted at Mach, who exercises civil and criminal powers. In 1903 the number of cognizable cases reported was fifteen, in *District staff, crime, and land revenue.*

seven of which convictions were obtained. The number of criminal cases was 45, and of civil cases 182. Land revenue at the rate of one-tenth of the produce was first levied in 1891, but the rate has since been raised to one-sixth. The land revenue in 1903–4 yielded Rs. 4,700, and the total revenue of the District from all sources was Rs. 9,500.

Local boards, police, and levies. A small sum is raised by a conservancy cess in the Mach bazar, and is spent on sanitation. In 1903–4 the receipts amounted to Rs. 1,100 and the expenditure to Rs. 900. The sanctioned strength of the levy force is 208 men, of whom 113 are employed in the pass, the remainder being detailed with the Political Agent, Kalāt, and elsewhere. The police force, which numbers thirty-nine men under two deputy-inspectors, is posted at eight railway stations and forms part of the Quetta-Pishīn police. No schools have been established. About twenty-seven pupils receive instruction in mosques.

Medical. A dispensary, maintained at Mach by the North-Western Railway, affords medical aid to the civil population. It has accommodation for thirteen in-patients. The total attendance in 1903 numbered 3,675. Vaccination has not been introduced.

Boundaries, configuration, and hill and river systems. **Sibi District** (*Sīwī*).—A District of Baluchistān, lying between 27° 55′ and 30° 38′ N. and 67° 17′ and 69° 50′ E. Its total area is 11,281 square miles; but this includes the MARRI-BUGTI country (7,129 square miles), which is only under political control, leaving 4,152 square miles of directly Administered territory. The Lahri *niābat* of the Kalāt State in Kachhi (1,282 square miles) is also politically controlled from Sibi. The District is bounded on the north by the Loralai District; on the south by the Upper Sind Frontier District; on the east by the Dera Ghāzi Khān District of the Punjab; and on the west by Kachhi, the Bolān Pass, and Quetta-Pishīn. The portion under political control occupies the centre, east, and south of the District; the areas under direct administration form protrusions in the north-western, north-eastern, and south-western corners.

No area in Baluchistān presents such strongly marked variations, both physical and climatic, between its various parts as the Sibi District. Two portions of it, the Sibi and Nasīrābād *tahsīls*, consist of perfectly level plain, lying respectively at the apex and base of Kachhi. The remainder of the District consists entirely of mountainous country, rising in a series of terraces from the lower hills of the SULAIMĀN range. These hills include Zen (3,625 feet) in the Bugti country, and Bambor (4,890 feet) and Dungān with Butur (about 6,000 feet) in the Marri country. North-westward the mountains stretch to the watershed of the CENTRAL BRĀHUI range in Zarghūn and Khalīfat, at an elevation of 11,700 feet. With the exception of the eastern side of the Marri-Bugti country, the drainage of the whole of this area is carried off by the NĀRI, which in traversing the Marri country is known as the Beji. On the south it is joined by three considerable hill-torrents, the Chākar or Talli, the Lahri, and the Chhatr. All of these streams are subject to high floods, especially in July and August, when the fertile lands of Kachhi are irrigated from them.

The upper, middle, and lower Siwāliks (upper and middle Geology. miocene); Spīntangi limestone and Ghāzij group (middle eocene); volcanic agglomerates and ash-beds of the Deccan trap; the Dunghān group (upper cretaceous); belemnite beds (neocomian); and some massive limestone (jurassic), as well as spreads of recent deposits, are exposed in the District.

The vegetation of the District is as varied as its physical Botany. aspects. On the south it is similar to that of Sind, the uncultivated land producing *Prosopis spicigera, Capparis aphylla, Salvadora oleoides, Zizyphus Nummularia, Tamarix indica, Acacia arabica,* and *Acacia modesta.* In the lower highlands the dwarf-palm (*Nannorhops Ritchieana*) abounds, and the blue gum (*Eucalyptus*) has been found to grow well. In the higher hills are found the juniper, pistachio, ash, wild almond, and *Caragana.* Cumin seed grows in the Ziārat hills, which also produce many varieties of grass.

Mountain sheep and *mārkhor* are found in the higher hills, Fauna. where leopards and black bears are also sometimes seen. Ravine-deer and hares occur in the plains. Large flocks of sand-grouse visit the District when there is a good mustard crop. Fair fishing is to be had in the Nāri.

While the highlands possess a climate which is pleasantly Climate, cool in summer and very cold in winter, the plains suffer from tempera-ture, and the great heat common in Sind. Nasīrābād has a mean tem-rainfall. perature in July of 96°, and is subject to the effects of the simoom. For five months alone, during the cold weather, are the climatic conditions tolerable to Europeans. The Marri-Bugti country and the Shāhrig *tahsīl* (2,300 to 4,000 feet) possess a climate intermediate between the extremes of the plains and the highlands. The annual rainfall varies with the altitude, from 3 inches in Nasīrābād to 5 in Sibi and nearly 12 in Shāhrig, where the vapour-bearing clouds strike Khalīfat and empty their contents into the valley.

History and archaeology.

Up to the end of the fifteenth century the District was always a dependency of Multān. It is known to have formed part of the Ghaznivid empire, and was ruled by a petty chief in the time of Nāsir-ud-dīn Kubācha. About 1500, it was taken by Shāh Beg, Arghūn, and thus passed under Kandahār, but, under the Mughal empire, it again became subordinate to Multān. It was taken by the Kalhoras of Sind in 1714; but they had to retire before the power of the Durrānis, by whom the local governors were generally selected from the Bārozai clan of the Panni Afghāns, which still retains much influence. During the last two years of the first Afghān War an Assistant Political Agent was posted to Sibi, and on its conclusion the District was handed over to Kalāt, but again came under Bārakzai rule in 1843. In the succeeding years the Marris acquired ground in the District; and their depredations were not checked until Sibi, Shāhrig, and Duki were assigned to the British, in 1879, by the Treaty of Gandamak. The Marris and Bugtis had been controlled from the Dera Ghāzi Khān District of the Punjab previous to the establishment of the Baluchistān Agency in 1877; and this charge now devolved on the Political Agent in Thal-Chotiāli, the name first given to the District on its establishment in 1879. The Kuat-Mandai valley, which belongs to the Marri tribe, has been held since 1881 as security for the payment of a fine inflicted after the Marri expedition of 1880. Owing to disputes between the Zarkūn Afghāns and the Marris, the Kohlu valley was brought under British protection in 1891. Nasīrābād was a *niābat* of the Kalāt State till 1903, when it was taken over on a perpetual lease for an annual payment of Rs. 1,15,000, increased by Rs. 2,500 in April, 1904. The name of the District was changed to Sibi in 1903, at which time the Sanjāwi, Duki, and Bārkhān *tahsīls*, which had hitherto formed part of the old Thal-Chotiāli District, were transferred to the new Loralai District.

The people, their tribes and occupations.

The Sibi District proper possesses one town and 304 villages, and its population in 1901 amounted to 73,893, or 18 persons per square mile. The Marri-Bugti country has eight villages and a population of 38,919. The total population, including tribal areas, is therefore 112,812. But this does not include the Dombkis (12,400), Umrānis (1,100), and Kaheris (7,100), who live in that portion of KACCHI which is controlled from the Sibi District. The table on the next page gives statistics of the area, &c., of the Administered territory by *tahsīls* in 1901.

In the Administered area 90 per cent. of the population are Muhammadans of the Sunni sect and 9 per cent. are

Hindus ; in the Marri-Bugti country the Muhammadans num-
ber 99 per cent. About 43 per cent. of the people speak
Baluchi ; the other languages spoken are Pashtū, Jatki, and
Sindī. A peculiar dialect, called Tarīno, is spoken in Shāhrig.
The Baloch number about 48,000 ; Afghāns follow with 18,000.
The Marris and Bugtis and the Dumars are large flock-owners ;
the other inhabitants are cultivators.

Tahsīl.	Area in square miles.	Number of		Population.	Population per square mile.
		Towns.	Villages.		
Kohlu . . .	362	...	9	1,743	5
Sibi . . .	1,343	1	32	20,526	15
Shāhrig . . .	1,595	...	93	16,573	10
Nasīrābād . .	852	...	170	35,713	42
Total	4,152	1	304	74,555*	81

* Includes 662 Marris enumerated in the Kohlu *tahsīl.*

The soil of the plains is alluvium, locally known as *pat* ; in the
lower highlands it is sandy ; in Kohlu it is much impregnated
with salt. Clay and gravel occur at the higher elevations. The
directly Administered area is well irrigated and fertile, but the
Marri and Bugti hills afford small opportunity for agriculture.
Of all the *tahsīls,* Kohlu alone has not been surveyed. The
total cultivable area in the remaining *tahsīls* is 878 square miles,
of which about 234 square miles are cultivated annually. The
principal harvest is the *sānwanri* or autumn crop ; wheat and
oilseeds compose the spring crop (*arhāri*). The largest area is
under *jowār,* after which come oilseeds and wheat. Rice,
millets, and gram are also grown. Cultivation has extended
everywhere with the advent of peace and security ; in Nasīrābād
it has risen from 76 square miles in 1880-1 to 165 square
miles in 1902-3, and in Sibi from about 7 square miles in
1879-80 to about 59 square miles in 1904. Quantities of
vegetables are raised in Sibi for the Quetta market, and the
cultivation of tobacco, potatoes, and melons is increasing.
Between 1897 and 1904 advances for agricultural improvements were given to the amount of nearly Rs. 50,000.

General agricultural conditions, statistics, and crops.

The class of cattle in the plains is excellent. The ponies of
the Marri and Bugti hills are light in limb and body, but carry
heavy weights unshod over the roughest ground. In the plains
larger animals are kept. The number of branded mares is 164.
Government stallions are stationed at Sibi in the winter.

Cattle, horses, sheep, &c.

Camels are bred in the southern part of the District. A horse and cattle fair is held at Sibi in February.

Irrigation. The Nasīrābād *tahsīl* is irrigated by the Desert and Begāri branches of the Government canals in Sind. The water is brought to the land either by gravitation (*moki*) or by lift (*charkhi*). The area irrigated annually between 1893 and 1903 averaged 80,000 acres. In the Sibi *tahsīl* a system of channels from the Nāri river irrigates about 26,000 acres. Elsewhere, excluding Kohlu, about 13,700 acres are irrigated from springs and streams. Wells are used for irrigation in Nasīrābād, but their number is limited. Most of the irrigated land is allowed to lie fallow for a year or two. The *kārez* number fourteen.

Forests. 'Reserved' juniper forests number seven, with an area of 69 square miles; and mixed forests, nine in number, cover about 41 square miles. The former are situated in Shāhrig, and seven of the latter are in the Sibi *tahsīl*. The juniper forests contain an undergrowth of wild almond (*Prunus eburnea*) and *mākhi* (*Caragana*); and the mixed forests grow *Prosopis spicigera*, *Capparis aphylla*, tamarisk, and acacia.

Minerals. Coal occurs in the Shāhrig *tahsīl*, and petroleum at Khattan in the Marri country. An account of the methods of working them will be found in the article on BALUCHISTĀN. The output of coal from Khost in 1903 amounted to 37,000 tons, but petroleum is no longer worked. An unsuccessful boring for oil was made in 1891 near Spīntangi. Earth-salt was manufactured in Nasīrābād up to 1902.

Arts and manufactures. Rough woollen fabrics, coarse carpets in the *dari* stitch, nose-bags, and saddle-bags are produced in many places. Felts and felt coats are made by the women of the highlands for domestic use. Mats, ropes, sacks, baskets, camel-pads, and many other articles are woven from the dwarf-palm, which is one of the most useful plants of the District. Embroidery is made by the Bugti women, the stitch chiefly used being herring-bone, with the threads looping through each other. The design often consists of large circular buttons or medallions joined by rings of chain stitch.

Commerce. The district produces *jowār*, wheat, *ghī*, and wool, and in years of good rainfall medicinal drugs, especially cumin seed, in some quantities. The only centre of trade is Sibi, the total imports and exports of which town by rail have risen from 11,800 tons in 1898 to 13,700 tons in 1903. Trade is largely carried on by agents of firms from Shikārpur in Sind. The principal imports into Sibi are gram, pulse, rice, dried fruits,

and piece-goods; the exports are *jowār, bājra,* wheat, and
oilseeds.

The Sind-Pishīn section of the North-Western Railway, on Railways
the standard gauge, enters the District near Jhatpat and, after and roads.
crossing the Kachhi plain, passes to Kach Kotal. Sibi is the
junction for the Mushkāf-Bolān branch. The centre and south
of the District are ill provided with roads. Partially metalled
roads extend to 125 and unmetalled tracks to 444 miles.
They are maintained chiefly from Provincial revenues and
partly from Local funds. The main routes consist of part
of the Harnai-Fort Sandeman road, and a cart-road from Sibi
to Kach and thence to Ziārat. A bridle path, which will form
an important artery, is now (1906) in course of construction
from Bābar Kach station to Kohlu via Māwand.

The Nasīrābād and Shāhrig *tahsīls* are fairly well protected Famine.
from famine owing to their extensive irrigation. Parts of the
Sibi and Kohlu *tahsīls* and of the Marri-Bugti country, however, depend almost entirely on rainfall, the failure of which
frequently results in scarcity. Between 1897 and 1901 the
rainfall was continuously deficient, and in 1897-8 about
Rs. 3,400 was expended in the Sibi *tahsīl* out of money
allotted by the Indian Famine Relief Fund. In 1899-1900
a sum of Rs. 18,000 was supplied from Imperial revenues for
grain doles to the Marris and Bugtis, and in the following
year Rs. 7,000 from the same source was distributed among
them for the purchase of bullocks and seed grain. A contribution of Rs. 6,459 from the Indian Famine Relief Fund was
also spent on the same objects in Sibi, Shāhrig, and Kohlu.
Between 1899 and 1901 District relief works cost about
Rs. 24,400.

The District consists of two portions: the Sibi District, con- District
taining the Sibi and Shāhrig *tahsīls*, which form part of British sub-
Baluchistān; and the Kohlu and Railway District, consisting and staff.
of the Kohlu and Nasīrābād *tahsīls* and the railway line lying
in Kachhi and the Marri country, which form part of the
Agency Territories. For purposes of administration the
District is treated as a single unit, in charge of a Political
Agent and Deputy Commissioner, with three subdivisions:
NASĪRĀBĀD, SIBI, and SHĀHRIG. Each of the first two is in
charge of an Extra Assistant Commissioner, and the latter of
the Assistant Political Agent. The Political Agent exercises
political control in the MARRI-BUGTI country, and over the
Dombki and Kaheri tribes of the Lahri *niābat* in Kachhi
through the Extra Assistant Commissioner at Sibi. Each

tahsīl has a *naib-tahsīldār*, except Kohlu, where a *naib-tahsīldār* exercises the powers of a *tahsīldār*. A Munsif is stationed at Sibi.

Civil justice and crime. The Deputy-Commissioner and Political Agent is the District and Sessions Judge. The Assistant Political Agent and the Extra Assistant Commissioners are magistrates of the first class, with power to try suits to the value of Rs. 10,000. *Tahsīldārs* are magistrates of the second class, with civil powers up to Rs. 300. *Naib-tahsīldārs* are magistrates of the third class, with civil powers in suits of the value of Rs. 50. The Munsif at Sibi is also a magistrate of the second class. Appeals from the officers of the lower grades lie to the sub-divisional officers. Many cases, in which the people of the country are concerned, are referred to *jirgas* for an award under the Frontier Crimes Regulation. The number of cognizable cases reported during 1903 was 134, convictions being obtained in 73 instances. The total number of criminal cases was 304 and of civil suits 1,209. The cases referred to *jirgas* numbered 645, including 17 cases of murder, 7 cases of robbery, 24 of adultery, and 15 cases of adultery accompanied by murder.

Land revenue administration. In Akbar's time Sibi was a *mahāl* of the Bhakkar Sarkār of the Multān *Sūbah*. It paid about Rs. 34,500, and furnished 500 cavalry and 1,500 infantry. The Panni tribe also supplied a separate contingent. Chhalgari, i. e. the Harnai valley, which depended on Kandahār, paid Rs. 240 in money, 415 *kharwārs* of grain, and supplied 200 horse and 300 foot. Under the Durrānis the revenue of the Sibi *tahsīl* was about Rs. 4,500. The present system of levying revenue varies in different parts of the District, and even in different areas within the same *tahsīl*. Fixed cash assessments, varying from Rs. 2 to Rs. 2–8–0 per acre on irrigated lands, are to be found side by side with the collection of an actual share of the produce (*batai*) at rates varying from one-fourth to one-twelfth. Details of each system are given in the separate articles on the *tahsīls* of the District. The annual value of the revenue-free holdings and grants of grain is Rs. 19,300. The land revenue, including grazing tax but excluding water-rate, amounted in 1903–4 to nearly 2 lakhs. This includes the revenue of Nasīrābād for six months only. The water-rate in Nasīrābād, amounting to 1·2 lakhs in 1903–4, is paid over to the Government of Bombay, as the Begāri and Desert Canals, which irrigate it, belong to the Sind system. The total revenue of the District from all sources was 2·4 lakhs in the same year.

The Sibi bazar fund and the Ziārat improvement fund are Local referred to in the articles on SIBI TOWN and ZIĀRAT. Octroi boards. and conservancy cess are levied in some bazars near the Sind-Pishīn railway, and are credited to the Shāhrig bazar fund, the money being spent on sanitary and other works under the direction of the Assistant Political Agent in charge of Shāhrig. The income in 1903–4 was Rs. 6,800, and the expenditure Rs. 6,300.

A small detachment of native infantry is stationed at Sibi. Army, The District Superintendent of Police at Quetta is in charge police, and of the regular police, which consisted, in 1904, of 199 con-jails. stables and 23 mounted men, under a European inspector and Honorary Assistant District Superintendent, with 6 deputy inspectors and 56 sergeants. It was distributed in twenty-four stations. The police employed on the railway line numbered 63. The total force of levies available amounts to 439 men, of whom 238 are mounted and 91 are em-ployed on the railway. These figures do not include 225 men stationed in the Marri-Bugti country, and 26 in the Lahri *niābat*. Local funds maintain twenty-one watchmen. There is a District jail at Sibi and four subsidiary jails, with total accommodation for 100 male and 24 female prisoners. Prisoners whose terms exceed six months are sent to the Shikārpur jail in Sind.

In 1904 the District had one middle and eight primary Education. schools, including a school for native girls and another for European and Eurasian boys and girls. The number of pupils was 342, and the annual cost Rs. 6,511, of which Rs. 2,284 was paid from Provincial revenues, and Rs. 4,187 from Local funds. The number of boys and girls receiving elementary instruction in mosque and other private schools was 926. Education in the Marri-Bugti country is represented by a single school at Dera Bugti.

The District possesses one hospital and four dispensaries, Hospitals, with accommodation for seventy-four patients. The average dispen-daily attendance of patients in 1903 was twenty-one. Two of saries, and the institutions are maintained by the North-Western Railway, tion. two are aided from Local funds, and the other is main-tained from Provincial revenues. The expenditure from Local funds and Provincial revenues in 1903 was Rs. 9,000. A female dispensary has recently been established at Sibi. Shāhrig has an evil reputation for malaria in summer, and syphilis is common in parts of the *tahsīl*. Malarial fever is the most prevalent disease throughout the District. Vaccina-

tion is optional and most of the people still resort to inoculation. The number of persons successfully vaccinated in
1903 was 3,363, or 46 per 1,000 on the total population of
the administered area.

[O. T. Duke: *Report on the District of Thal-Chotiāli and
Harnai.* (Foreign Department Press, 1883.)—R. I. Bruce:
*History of the Marri Baloch Tribe and its Relations with
the Bugti Tribe.* (Lahore, 1884.)—*Bombay Records*, No.
XVII, New Series, containing, among other papers, a Diary
kept by Captain Lewis Brown while besieged in Kahān.—
R. D. Oldham: 'Geology of Thal-Chotiāli aṇd part of the
Marri Country,' *Records Geol. Survey of India*, vol. xxv, part 1.
—C. L. Griesbach: 'Geology of the Country between the
Chappar Rift and Harnai,' ib. vol. xxvi, part 4.]

Sibi Subdivision.—A subdivision of the Sibi District,
Baluchistān, comprising the *tahsīls* of SIBI and KOHLU. The
Extra Assistant Commissioner in charge also exercises political
control in the Marri-Bugti country and in the Lahri *niābat* of
the Kalāt State in Kachhi.

Kohlu.—A *tahsīl* of the Sibi subdivision in the District of
the same name, Baluchistān, lying between 29° 43′ and 30° 2′
N. and 68° 58′ and 69° 32′ E. Its area is 362 square miles,
and population (1901) 1,743. It forms a triangular plateau
about 3,900 feet above sea-level and has a pleasant climate.
The head-quarters bear the same name as the *tahsīl*. Villages
number nine. The land revenue in 1903–4 amounted to Rs.
14,154. On lands acquired by the Marris previous to 1892
revenue at the rate of one-twelfth of the produce is taken, an
equal share being paid by the cultivator to the Marri chief. On
other lands revenue is levied at the rate of one-sixth.

Sibi Tahsīl (*Sīwi*).—A *tahsīl* of the Sibi District, Baluchistān, lying between 29° 21′ and 30° 15′ N. and 67° 11′ and 68°
9′ E., at the apex of the Kachhi plain, and including the hilly
country round Sāngān. It has an area of 1,343 square miles,
and a population (1901) of 20,526, showing an increase of
7,125 since 1891. It possesses one town of importance, SIBI
(population, 4,551), and thirty-two villages. The land revenue in
1903–4 amounted to 1·1 lakhs. The rate of revenue levied in
Sibi is two-ninths of the produce, as distinguished from the
usual one-sixth; in Sāngān it is one-fourth, half of which is paid
over to the Bārozai chief, and in Kuat-Mandai one-twelfth, the
Marri chief taking an equal amount. The *tahsīl* is irrigated
by canals from the Nāri river.

Shāhrig.—A subdivision and *tahsīl* of the Sibi District,

Baluchistān, lying between 29° 49' and 30° 37' N. and 67° 14' and 68° 22' E. Its area is 1,595 square miles, and population (1901) 16,573, showing an increase of only 332 since 1891. The head-quarters station is Shāhrig, but the Assistant Political Agent in charge of the subdivision generally resides at Ziārat or Sibi. The *tahsīl* possesses ninety-three villages. The land revenue, including grazing tax, was Rs. 28,900 in 1903–4. All irrigated lands are under a fixed cash assessment for a term of ten years, which terminates in 1911. The incidence per irrigated acre ranges from Rs. 2–14–11 to Rs. 2–2–6. Besides the Zāwar or Harnai valley, the *tahsīl* includes a mass of mountainous country on the north, intersected by the picturesque Kach-Kawās valley leading to ZIĀRAT. It possesses the distinction of having the highest recorded rainfall in Baluchistān (11·67 inches).

Nasīrābād.—A subdivision and *tahsīl* of the Sibi District, Baluchistān, lying between 27° 55' and 28° 40' N. and 67° 40' and 69° 20' E., on the border of the Upper Sind Frontier District of Sind. It has an area of 852 square miles and a population (1901) of 35,713, and, for administrative purposes, includes the railway line from the neighbourhood of Jhatpat to Mithri. The head-quarters of the *tahsīl* are at present (1906) at Nasīrābād, about 8 miles from Jacobābād. It contains 170 villages. It depends for cultivation on the Begāri and Desert Canals of the Sind system, and is the only *tahsīl* in Administered territory in which indigo and gram are produced. In 1904–5, the first complete year of administration, the land revenue, excluding water-rate, amounted to 1·2 lakhs. Water-rate is levied at R. 1 per irrigated acre on the Begāri Canal, and at Rs. 1–8–0 on the Desert Canal. The incidence of land revenue is R. 1 per acre, and a special cess of 6 pies is also collected. A revision of the rates is contemplated, beginning from 1905.

Marri-Bugti Country.—A tribal area in Baluchistān, controlled from the Sibi District, lying between 28° 26' and 30° 4' N. and 67° 55' and 69° 48' E., with an area of 7,129 square miles. The northern part, the area of which is 3,268 square miles, is occupied by the Marris, and the southern part, 3,861 square miles, by the Bugtis. The country is situated at the southern end of the Sulaimān range. It is hilly, barren, and inhospitable, and supplies are scarce. Here and there are good pasture grounds, and a few valleys and plains are gradually being brought under cultivation. The valleys and plateaux include Nisau (3,000 feet), Jant Alī (2,847 feet), Kahān (2,353 feet), Māwand (2,620 feet), and Marav (2,195 feet). The rainfall is scanty and is chiefly received in July.

The Marris and Bugtis are the strongest Baloch tribes in the Province. The total population of their hills was 38,919 in 1901, or about five persons to the square mile. The Marris, including those living in the British *tahsīl* of Kohlu, numbered 19,161, with 140 Hindus and 1,090 other persons living under their protection (*hamsāyah*). The population of the Bugti country amounted to 18,528, comprising 15,159 Bugtis, 272 Hindus, 708 *hamsāyahs*, and 2,389 *maretās* or servile dependants. The whole population is essentially nomadic in its habits, and lives in mat huts. The total number of permanent villages in the country decreased from eight in 1901 to five in 1904; the most important are Kahān (population about 400) in the Marri country, and Dera Bugti (population about 1,500) in the Bugti country.

Both tribes are organized on a system suitable to the predatory transactions in which they were generally engaged in former times. Starting from a small nucleus, each gradually continued to absorb various elements, often of alien origin, which participated in the common good and ill, until a time arrived when it was found necessary to divide the overgrown bulk of the tribe into clans (*takkar*), the clans into sections (*phalli*), and the sections into sub-sections (*pāra* or *firka*). At the head of the tribe is the chief (*tumandār*), with whom are associated the heads of clans (*mukaddam*) as a consultative council. Each section has its *wadera*, with whom is associated a *mukaddam*, who acts as the *wadera's* executive officer and communicates with the *motabars* or headmen of sub-sections. Each tribe was thus completely equipped for taking the offensive. In pre-British days a share of all plunder, known as *panjoth*, was set aside for the chief; headmen of clans then received their portion, and the remainder was divided among those who had taken part in an expedition. Side by side with this system there still exists, among the Marris and the Pairozāni Nothāni clan of the Bugtis, a system of periodical division of all tribal land. The three important clans of the Marris are the Gaznis (8,100), to whom the Bahāwalānzai or chief's section belongs; the Loharāni-Shirāni (6,400); and the Bijrāni (4,700). The Bugtis include the clans of Pairozāni Nothāni (4,700), Durragh Nothāni (1,800), Khalpar (1,500), Massori (2,900), Mondrāni (500), Shambāni (2,900), and Raheja (880). The chief's section belongs to the latter. The chiefs levy no revenue, but usually receive a sheep or a goat from each flock when visiting different parts of their country.

The early history of both tribes is obscure. The Marris are known to have driven out the Kupchānis and Hasnis, while the Bugtis conquered the Buledis. Owing to the great poverty of their country, both tribes were continuously engaged in plunder and carried their predatory expeditions far into the adjoining regions. They came in contact with the British during the first Afghān War, when a force under Major Billamore penetrated their hills. In April, 1840, a small detachment was sent, under Captain Lewis Brown, to occupy Kahān and guard the flank of the lines of communication with Afghānistān; but it was invested for five months and two attempts at relief were beaten off. The fort was, however, only surrendered after a safe retreat had been secured from Doda Khān, the Marri chief. In 1845 Sir Charles Napier conducted a campaign against the Bugtis, who fled to the Khetrāns, and the expedition was only a qualified success. General John Jacob, after much trouble with both tribes, but especially with the Bugtis, settled some of the latter on irrigated lands in Sind in 1847, but many of them shortly afterwards fled to their native hills. Both tribes were subsidized by the Khān of Kalāt after the treaty of 1854; but in 1859 Mīr Khudādād Khān was obliged to make an expedition against the Marris, accompanied by Major (afterwards Sir Henry) Green. Another unsuccessful campaign followed in 1862. Anarchy ensued; and in 1867 Captain (afterwards Sir Robert) Sandeman, the Deputy-Commissioner of Dera Ghāzi Khān, entered into direct relations with them and took some of them into the service of Government. The result of the Mithankot conference, which took place between Punjab and Sind officials in 1871, was to place Sandeman in political control of the Marri-Bugti country under the orders of the Superintendent, Upper Sind Frontier.

On the establishment of the Baluchistān Agency in 1877, British relations with the Marris and Bugtis became closer, and service and allowances were given to them. The Bugtis have throughout behaved well. The Marris, in August, 1880, plundered a convoy marching along the Harnai route and killed forty-two men, whereupon a punitive expedition was dispatched under General Sir Charles Macgregor, to whom the Marri chief and his headmen tendered their submission. They paid Rs. 1,25,000 in cash, out of a fine of Rs. 1,75,000 inflicted on them, and agreed to surrender half of the revenue of the Kuat-Mandai valley until the balance of Rs. 50,000 had been paid off. Since then the Marris have given little trouble, with the exception of the part they took in the Sunari outrage in 1896, when they killed eleven men, and some unrest which occurred in 1898 and ended in the son of the Marri chief emigrating temporarily to Afghānistān.

Both tribes are under the control of the Political Agent in Sibi, with the Extra Assistant Commissioner of the Sibi subdivision in subordinate charge. Direct interference in the internal affairs of the tribes is, so far as possible, avoided, the chiefs being left to decide all such cases in consultation with their sectional headmen and in accordance with tribal custom. The task of the Political officers is chiefly confined to the settlement of intertribal cases either between the Marris and Bugtis themselves, whose relations are frequently strained, or with the neighbouring tribes of the Loralai District and the Punjab. A code of penalties for the infliction of particular injuries, such as murder, the loss of an eye or tooth, &c., was drawn up between the Marris and Bugtis in 1897 and is followed in ordinary circumstances. Cases of extraordinary importance are referred to the *Shāhi jirga*, and the Political Agent sees that the award is carried out. Large services have been given to both tribes to enable the chiefs to secure control over their followers. The Marri tribal service consists of 1 headman, 206 mounted levies, 5 footmen, and 8 clerks and menials; 35 of these men are stationed in seven posts in the Loralai District and 109 at thirteen posts in the Administered area of Sibi District. The remainder hold three posts in the Marri country. The total monthly cost amounts to Rs. 5,600. The Bugti service includes 3 headmen, 136 mounted levies, 4 footmen, and 6 clerks, costing Rs. 3,800 monthly. The posts on the south of the Bugti country are controlled from the Nasīrābād *tahsīl*.

Thal-Chotiāli.—A former District of Baluchistān, the north-eastern part of which has been merged since 1903 in the LORALAI District and the southern and western parts in the SIBI District.

Harnai.—A railway station and village in Baluchistān, on the Sind-Pishīn section of the North-Western Railway, situated in 30° 6′ N. and 67° 56′ E., at an elevation of 3,000 feet. It lies in the valley of the same name in the Shāhrig *tahsīl* of the Sibi District, and is the starting-point of the road for LORALAI (55¾ miles) and for FORT SANDEMAN (168 miles), with which it is connected by a cart-road. Harnai contains a small bazar, police station, dispensary, and dāk-bungalow.

Sibi Town (*Sīwi*).—Head-quarters of the Sibi District, Baluchistān, situated in the *tahsīl* of the same name, in 29° 33′

N. and 67° 53′ E., 88 miles from Quetta and 448 from Karāchi. The population numbered 4,551 in 1901, an increase of 1,607 since 1891. The place is very old and is mentioned as early as the thirteenth century. Owing to its exposed situation, between the mouths of the Harnai and Bolān Passes, it has suffered from frequent sieges, including an assault by the British in 1841. The existing town dates from 1878. It possesses a considerable trade. The Victoria Memorial Hall, erected by public subscription in 1903, is the only building of importance. A piped water-supply has been provided by military funds from the Nāri river at a cost of Rs. 1,15,000. Though not a municipality, a town fund is maintained, the income of which in 1903–4 amounted to Rs. 23,700 and the expenditure to Rs. 23,000.

Ziārat.—A sanitarium and the Provincial summer headquarters of the Baluchistān Agency, situated in 30° 23′ N. and 67° 51′ E., at an elevation of 8,050 feet above the sea. It lies in the Shāhrig *tahsīl* of the Sibi District, and is the residence of the Political Agent from May to October. Ziārat is most easily reached from Kach station on the North-Western Railway by a cart road (32¾ miles). The local name is Gwashki, which was changed in 1886 to Ziārat, after the neighbouring shrine of Mīan Abdul Hakīm. It was first visited and selected as a sanitarium in 1883. The Residency was built in 1890–1. The climate during the short summer is delightful, and the air is bracing. A piped water-supply was provided in 1898–9 at a cost of Rs. 38,000. The hill-sides are covered with juniper, and there are many lovely walks through the wooded glades. Huge gorges and defiles constitute a feature of the scenery. Besides the Residency, the remaining edifices consist of houses for officials and other Government buildings. Sanitation is provided for by the Ziārat improvement fund, a branch of the Shāhrig bazar fund. The income in 1903–4 amounted to Rs. 3,800 and the expenditure to Rs. 2,689. A summer camp for the European troops stationed at Quetta was first formed at Ziārat in 1885, but the experiment was afterwards abandoned until 1903, since which year the camp has again been established. It is located on a spur of the Batsarg hill.

Kalāt State.—A Native State in Baluchistān, lying between 25° 1′ and 30° 8′ N. and 61° 37′ and 69° 22′ E., with a total area of 71,593 square miles. It occupies the whole of the centre and south-west of the Province, with the exception of the indentation caused by the little State of Las Bela. It is bounded on the west by Persia; on the east by the Bolān Pass, the Marri and Bugti hills, and Sind; on the north by the *Boundaries, configuration, and hill and river systems.*

Chāgai and Quetta-Pishīn Districts ; and on the south by Las
Bela and the Arabian Sea. With the exception of the plains of
Khārān, Kachhi, and Dasht in Makrān, the country is wholly
mountainous, the ranges being intersected here and there by
long narrow valleys. The principal mountains are the CENTRAL
BRĀHUI, KĪRTHAR, PAB, SIĀHĀN, CENTRAL MAKRĀN, and
MAKRĀN COAST ranges, which descend in elevation from
about 10,000 to 1,200 feet. The drainage of the country is
almost all carried off to the southward by the NĀRI, MŪLA,
HAB, PORĀLI, HINGOL, and DASHT rivers. The only large
river draining northwards is the RAKHSHĀN. The coast-line
stretches for about 160 miles, from near Kalmat to Gwetter
Bay, and the chief port is PASNI. Round GWĀDAR the country
is in the possession of the Sultān of Maskat.

Geology.   The geological groups in the State include liassic ; jurassic
(lower and upper cretaceous strata) ; volcanic rocks of the
Deccan trap ; Kīrthar (middle eocene) ; lower Nāri (upper
eocene ; and Siwālik beds (middle and upper miocene), besides
extensive sub-recent and recent deposits.   The State also
includes a portion of the Indus alluvial plain.

Botany.   The botany of the north differs entirely from that of the
south. In the former the hill slopes occasionally bear juniper,
olive, and pistachio ; poplars, willows, and fruit trees grow in
the valleys ; herbaceous and bulbous plants are frequent on the
hill-sides ; and in the valleys southernwood (*Artemisia*) and
many *Astragali* occur. In the latter the vegetation consists of
a thorny unpleasant scrub, such plants as *Capparis aphylla*,
Prosopis spicigera, *Calotropis procera*, *Acanthodium spicatum*,
and *Acacia* being common. The dwarf-palm (*Nannorhops
Ritchieana*) affords a means of livelihood to many of the
inhabitants.

Fauna.   Sind ibex and mountain sheep occur, but are decreasing in
numbers. Ravine-deer are common. Bears and leopards are
found occasionally. *Sīsī* and *chikor* are abundant in the higher
hills. The wild ass is found in the western desert.

Climate,   The climatic conditions vary greatly. Along the coast con-
tempera-   ditions are intermediate between those of India and the
ture, and
rainfall.   Persian Gulf. Farther inland great heat is experienced during
summer, and the cold weather is short. Kachhi is one of
the hottest parts of India. Round Kalāt, on the other hand,
the seasons are as well marked as in Europe ; the temperature
in summer is moderate, while in winter severe cold is experienced and snow falls. All the northern parts depend on the
winter snow and rain for cultivation ; in the south most of the

rain falls in summer; everywhere it is irregular, scanty, and local.

History and archaeology.
The history of the State has been given in the historical portion of the article on BALUCHISTĀN. After being held successively by Sind, by the Arabs, Ghaznivids, Ghorids, and Mongols, and again returning to Sind in the days of the Sūmras and Sammas, it fell under the Mughal emperors of Delhi. The Ahmadzai power rose in the fifteenth century and reached its zenith in the eighteenth, but it was always subject to the suzerainty of Delhi or Kandahār. After the first Afghān War Kalāt came under the control of the British—a control which was defined and extended by the treaties of 1854 and 1876.

The most interesting archaeological remains in the country are the Kausi and Khusravi *kārez* in Makrān, and the ubiquitous stone dams known as *gabrbands* or 'embankments of the fire-worshippers.' Mounds containing pottery are frequent, and Buddhist remains have been found in Kachhi.

KALĀT TOWN is the capital of the State. Other towns of importance are Bhāg, Gandāva, Mastung, PASNI, and GWĀDAR. Permanent villages number 1,348 or 1 to 53 square miles; the majority of the population live in mat huts or in blanket tents.

The people, their tribes and occupations.
The State is divided into five main divisions: Kachhi, Sarawān, Jhalawān, Makrān, and Khārān, the latter being quasi-independent. The population, which numbers (1903) 470,336, consists chiefly of Brāhuis and Baloch, but also includes Jats, who are cultivators in Kachhi; Darzādas and Nakībs, the cultivating class of Makrān; Loris, who are artisans; Meds and Koras, who are fishermen and seamen; and servile dependants. The traders consist of Hindus and a few Khojas on the coast. The majority of the people are Sunni Muhammadans, but, in the west, many belong to the sect called Zikri. Except in Makrān and Khārān, the people are organized into tribes, each of which acknowledges the leadership of a chief. Besides these tribesmen, who form the BRĀHUI confederacy with the Khān of Kalāt at its head, a distinct body is found in the Khān's own *ulus* or following, consisting of the cultivators in those portions of the country from which the Khān collects revenue direct. They are chiefly Dehwārs and Jats. Agriculture, flock-owning combined with harvesting, and fishing constitute the means of livelihood of most of the population. Brāhui, Baluchi, Dehwāri, and Sindī are the languages chiefly spoken.

Agriculture.
The soil is sandy in most places; here and there alluvial

deposits occur and a bright red clay, which gives place in Makrān to the white clay known as *milk*. Permanent irrigation is possible only in a few favoured tracts; elsewhere, the country depends almost entirely on flood cultivation from embankments. In irrigated tracts the supply of water is obtained from *kārez*, springs, and rivers. The staple food-grains consist of wheat and *jowār*. In Makrān the date is largely consumed. Rice, barley, melons, millets, tobacco, lucerne, potatoes, and beans are also cultivated. The commonest tree in the orchards is the pomegranate; and apricots, almonds, mulberries, vines, and apples are also grown. Experiments in sericulture are being made at Mastung (1906).

Cattle, horses, sheep, and fishing. An excellent breed of cattle comes from Nāri in Kachhi. The Sarawān country and Kachhi produce the best horses in Baluchistān. Kalāt possessed 783 branded mares in 1904. Large donkeys are bred near Kalāt town, and those in Makrān are noted for their speed. Sheep and goats are very numerous. The sheep's wool, of which large quantities are exported, is coarse and comes into the market in a deplorable condition of dirt. The goats are generally black. Camels are bred in large numbers in Kachhi, the Pab hills, and Khārān, and animals for transport are available almost everywhere. All households keep fowls. The better classes breed good greyhounds for coursing. The fishing industry on the Makrān coast is important and capable of development. Air-bladders, shark-fins, and salt fish are exported in large quantities.

Material condition of the people. Very little money circulates in the country, both rents and wages being usually paid in kind, and most of the tribesmen's dealings are carried on by barter. Owing to the inhospitable nature of the country, the people are very poor. The standard of living has risen slightly of recent years, and the people are now better clothed than formerly. A Brāhui will never beg in his own country. With the Makrānis mendicancy, which is known as *pindag*, is extremely common.

Forests and minerals. No arrangements for forest 'reservation' exist in the State; here and there, however, tribal groups preserve special grounds for grass and pasturage. Among minor forest products may be mentioned cumin seed, asafoetida, medicinal drugs, the fruit of the pistachio, bdellium, and gum-arabic. Few minerals have been discovered, and only coal, which occurs in the Sor range in the Sarawān country, is systematically worked. Traces of coal have been found elsewhere in the Sarawān country. Ferrous sulphate is obtainable in the Jhalawān country, and lead was at one time worked at Sekrān in the same area. Good earth-salt is obtainable from the swamps, known as _hāmūn_ or _kap_, and is also manufactured by lixiviation.

Coarse cotton cloth is woven in Kachhi and articles of floss silk are made in Makrān. All Brāhui women are expert with the needle, and the local embroidery is both fine and artistic. Rugs, nose-bags, &c., woven by nomads in the _darī_ stitch, are in general use. The art of making pile-carpets is known here and there. Durable overcoats (_shāi_) are made by the women from dark sheep's wool. Leather is embroidered in Kachhi, Kalāt, and Mastung. Matting, bags, ropes, and other articles are manufactured from the dwarf-palm. *Arts and manufactures.*

Commerce is hampered by the levy of transit-dues and octroi, both by the State and by tribal chiefs, and by the expense of camel-transport. The chief centres of trade are Kalāt, Mastung, Gandāva, Bhāg, Turbat, Gwādar, Pasni, and Nāl. The exports consist of wool, _ghī_, raw cotton, dates, salt fish, matting, medicinal drugs, and cattle, in return for which grain, piece-goods, metals, and silk are imported. From the north the traffic goes to Quetta; from the centre to Kachhi and Sind; and from the south and west by sea and land to Karāchi. *Commerce.*

The North-Western Railway traverses the east and north-east of the State. The only cart-road is that from Quetta to Kalāt town. All other communications consist of tracks for pack-animals, the most important of which are those connecting Kalāt with Panjgūr, Kalāt with Bela via Wad, and Kachhi with Makrān via the Mūla Pass. A track is now (1906) in course of construction from Pasni on the coast to Panjgūr. A postal service to Kalāt is maintained by the British Government, and letters are carried thence once a week to Khuzdār. The British India Company's mail steamers touch at Pasni and Gwādar on alternate weeks, and mails are carried from Pasni to Turbat, the head-quarters of Makrān. The Indo-European Telegraph wire traverses the coast, with offices at Pasni and Gwādar; a telegraph line runs from Quetta to Kalāt, and a line has been sanctioned from Karāchi to Panjgūr. *Communications.*

The State experiences constant scarcity and occasional famine. A drought lasting for ten years between 1830 and 1840 is mentioned by Masson. The population is, however, sparse and exceedingly hardy, and they have ready access to Sind, where good wages are obtainable. In the Census of 1901 so many as 47,345 Brāhuis were enumerated there. Advances amounting to about Rs. 29,000 were made by the State in 1900, when the scarcity which had begun in 1897 *Famine.*

reached its culminating point. Such advances are recovered from the cultivator's grain-heap at the ensuing harvests.

System of adminis-tration. The control exercised by the British Government over the Brāhui confederacy, and the administrative arrangements in areas subject to the direct authority of the Khān of Kalāt, are described in the article on BALUCHISTĀN. Except Khārān and Makrān, each main division of the State comprises both tribal areas and areas subject solely to the Khān. Collateral authority is, therefore, exercised by the Khān in his *niābats* and by tribal chiefs in their country. The intervention of the Political Agent is confined, so far as possible, to deciding inter-tribal cases or cases between the tribesmen and the Khān's subjects in which a right of arbitration rests with the British Government. In Makrān, the Khān's *nāzim* exercises authority everywhere; in Khārān, the chief is now subject to no inter-ference from the Khān, but looks to the Political Agent in Kalāt. The Quetta, Nushki, and Nasīrābād *tahsīls* have been leased in perpetuity by the State to the British Govern-ment, and the right to levy transit dues in the Bolān Pass has been commuted for an annual subsidy of Rs. 30,000. The head-quarters of the Political Agent were fixed at Mastung in 1904.

Finance. The revenue of the State is derived from three principal sources : subsidies and rents paid by the British Government, interest on investments, and land revenue. The subsidies include Rs. 1,00,000 paid under the treaty of 1876 and Rs. 30,000 for the Bolān Pass, while the quit-rents for the leased areas mentioned above amount to Rs. 1,51,500. Since 1893 a surplus of 41·5 lakhs has been invested in Govern-ment securities, yielding in interest 1·5 lakhs per annum. From this source are defrayed the cost of maintenance of the former Khān, Mīr Khudādād, the subsidies paid to the Jhalawān chiefs, the pay of Brāhui *thānas*, and the expenses of the administration of Makrān. The total income of the State may be estimated at between $7\frac{1}{2}$ and $8\frac{3}{4}$ lakhs of rupees, the variations being due to fluctuations in the land revenue. The expenditure amounts to about $3\frac{1}{2}$ or 4 lakhs. A sum of Rs. 53,000 is expended annually in the State by the British Government, in the shape of telegraph subsidies, payments to chiefs for controlling their tribesmen, and the maintenance of levies. To this will now be added the charges, amounting to about 1·2 lakhs per annum, for the Makrān Levy Corps.

Land revenue. Land revenue is collected in kind, the rates varying from one-third to one-eighth of the produce. Cesses are also taken, the amount of which differs in almost every village, but which raise the share taken by the State to nearly one-half. Here and there are to be found cash assessments (*zar-i-kalang* or *zar-i-shāh*). The cultivators also perform certain services for the Khān, such as the escort of his horses and the repairs to the walls of his forts. Transit-dues (*muhāri*) are levied on caravans passing through the *niābats*, and octroi (*sung*) on their entering and leaving trading centres. Contracts are given for the sale of liquor, meat, &c. The total land revenue varies with the agricultural conditions of the year. In 1903–4, on the introduction of a new system of administration, it rose to 4·5 lakhs. Large areas are held by tribesmen and tribal chiefs, in which the Khān is entitled to no revenue. In others, half the revenue has been alienated by the Khān (*adh-ambāri*). Many of these *jāgīrs* were originally held on the condition of feudal service. In Makrān the Gichkis, Nausherwānis, Bīzanjaus, and Mīrwāris are the principal holders, while in Kachhi the *jāgīrs* are held by Brāhuis and Baloch. In such areas the tribal chiefs claim complete independence in all revenue, civil, and criminal matters. In *adh-ambāri* areas the Khān retains jurisdiction.

The army is an irregular force, without organization or *Army.* discipline, consisting of 300 infantry, 300 cavalry, and 90 artillery with twenty-nine old-fashioned guns, of which none are serviceable. The infantry is divided into two regiments, and the cavalry into three. The total cost amounts to about Rs. 82,000 per annum. Most of the troops are at Kalāt; detachments are stationed at Mastung and Khuzdār, and in Kachhi. Sepoys are paid Rs. 6 a month; non-commissioned officers Rs. 7 to Rs. 12; while *risāldārs* and commandants receive from Rs. 20 to Rs. 50. The cavalry soldiers are mounted on horses found by the State. A force of 160 men is also maintained in Makrān, at an annual cost of about Rs. 32,000. Between 1894 and 1898 a body of 205 infantry and 65 camelmen under a British officer, known as the Kalāt State Troops, was maintained, but has been disbanded.

At the most important places in the Khān's *niābats* levies, *Levies,* known as *amla*, are stationed. These men are used for all *police, and* kinds of duties, both revenue and criminal. They number *jails.* 222, of whom 118 are mounted on their own horses and 64 are supplied with horses, when required, by the Khān. The remainder are unmounted. They are paid in kind, and get Rs. 18 per annum in cash. The total cash payments made to them amount to about Rs. 4,000. For dealing with cases in

which Brāhuis are concerned, *thānas*, manned by Brāhui tribesmen, are located in different parts of the country. They
number eleven, with 100 men. In tribal areas and *jāgīrs* the
peace is maintained by the chiefs, subsidies amounting to
about Rs. 50,000 being paid by the Khān for this purpose in
addition to the amounts paid by the British Government.
A force of ten police is attached to the Political Adviser to the
Khān for escort duty. One jail is maintained, with accommodation for 100 prisoners, and there are lock-ups at the Brāhui
thānas. Offenders are often kept in the stocks, and are fed by
their relations.

Education, Education has hitherto been entirely neglected, but a large
medical, school is about to be opened at Mastung. A few boys are
and sur-
veys. taught in mosque schools, and Hindu children receive education
from their parents. Two dispensaries are maintained, one by
the British Government and the other by the State. They
relieved 8,919 patients in 1903 and cost Rs. 5,300. Inoculation is practised everywhere, principally by the Saiyids and
Shaikhs, but the people have no objection to vaccination. The
whole country has been surveyed on the $\frac{1}{2}$-inch scale up to
66° E.; westward the results of a reconnaissance survey have
been published on the $\frac{3}{4}$-inch scale.

[*Baluchistān Blue Books,* Nos. 1, 2, and 3 (1887).—H. Pottinger: *Travels in Beloochistan and Sinde* (1816).—C. Masson: *Narrative of a Journey to Kalāt* (1843); *Journeys in
Baluchistān, Afghānistān, and the Punjab* (1842).—G. P. Tate:
Kalāt. (Calcutta, 1896.)]

Boun- **Sarawān.**—The northern of the two great highland divisions
daries, of the Kalāt State, Baluchistān, as distinguished from the
configura-
tion, and southern or Jhalawān division. It lies between 28° 57′ and
hill and 30° 8′ N. and 66° 14′ and 67° 31′ E., and is bounded on the
river
systems. east by Kachhi; on the west by the Garr hills, a continuation
of the Khwāja Amrān; on the north by the Quetta-Pishīn,
Bolān Pass, and Sibi Districts; and on the south by the
Jhalawān country. The total area of the country is 4,339
square miles. It consists of a series of parallel mountain
ranges running north and south and enclosing valleys, sometimes of considerable extent, which lie at an elevation of from
5,000 to 6,500 feet above sea-level. Reckoning from east to
west, the principal mountain ranges are the Nāgau, Bhaur, and
Zāmuri hills, which border on Kachhi; and the Bangulzai hills,
with the peaks of Moro and Dilband. Southward of these lies
the fine Harboi range, about 9,000 feet high. Westward again
the Koh-i-mārān (10,730 feet) forms another parallel ridge.

Next, the Zahri-ghat ridge commences from the Chiltan hill and skirts the Mastung valley to the east, while two more minor ranges separate it from the westernmost ridge, the Garr hills. Most of these mountains are bleak, bare, and barren, but the Harboi and Koh-i-mārān contain juniper trees and some picturesque scenery. The drainage of the country is carried off northward by the Shīrīnāb and Sarawān rivers. Except in flood time, each contains only a small supply of water, disappearing and reappearing throughout its course. The Shīrīnāb rises to the south-east of Kalāt. It is joined by the Mobi and Gurgīna streams, and eventually falls into the PISHĪN LORA under the name of the Shorarūd or Shar-rod. The Sarawān river rises in the Harboi hills and joins the Bolān near Bībī Nāni.

The principal peaks of the country consist of massive lime- Geology, stone; and cretaceous beds of dark, white, and variegated botany, and limestone, sometimes compact, sometimes shaly in character, fauna. occur. Sandstones, clays, and conglomerates of Siwālik nature have also been found. The botany of Sarawān resembles that of the Quetta-Pishīn District. Orchards, containing mulberry, apricots, peaches, pears, apples, almonds, and grapes, abound in the valleys. Poplars and willows grow wherever there is water, and tamarisk is abundant in the river-beds. In the spring many plants of a bulbous nature appear, including tulips and irises. The hill-sides are covered with southernwood (*Artemisia*) and many species of *Astragali*. Mountain sheep and Sind ibex occur in small numbers. Foxes are trapped for their skins, and hares afford coursing to local sportsmen.

From April to September the climate is dry, bright, bracing, Climate, and healthy. The winter, especially round Kalāt, which receives temperature, and heavy falls of snow, is severe. Except on the east, near Bārari, rainfall. the heat in summer is nowhere intense. The rain and snowfall generally occur in winter, from January to March. The average annual rainfall is about 7½ inches, of which 6 inches are received in winter and 1½ in summer.

The Sarawān country formed part of the Ghaznivid and History Ghorid empires, and fell into the hands of the Arghūns towards and the end of the fifteenth century. From them it passed to the logy. Mughals until, towards the end of the seventeenth century, Mīr Ahmad of Kalāt acquired Mastung from Aghā Jāfar, the Mughal governor. Henceforth Mastung remained under Kalāt and was the scene of an engagement between Ahmad Shāh Durrāni and Nasīr Khān I in 1758, in which the Afghāns were at first defeated, but Ahmad Shāh afterwards advanced and assaulted Kalāt. During the first Afghān War, the country was one of the districts assigned by the British in 1840 to Shāh Shujā-ul-mulk, but it was restored to Kalāt in 1842. During 1840 the Sarawān tribesmen revolted and placed Nasīr Khān II on the throne. In 1871 another rebellion occurred, and the Brāhuis received a crushing defeat from Mīr Khudādād Khān at Khad near Mastung. In 1876 the latter place was the scene of the memorable settlement effected by Sir Robert Sandeman between Khudādād Khān and his rebellious chiefs.

Curious mounds situated in the centre of the valleys occur throughout the country. Two of the largest are Sāmpūr in Mastung and Karbukha in Mungachar. They are artificial, being composed of layers of soil, ashes, and broken pottery.

The people, their tribes and occupations.

KALĀT TOWN, and Mastung, the head-quarters of the Political Agent, are the only towns. The country possesses 298 permanent villages. The population in 1901 was 65,549. Most of the people make their way to Kachhi in the winter. The centre of the country is inhabited by the cultivating classes known as Dehwārs, Khorāsānis, and Johānis, most of whom are subjects of the Khān of Kalāt. In the surrounding hills and vales live the tribesmen composing the Sarawān division of the Brāhui confederacy. They include the Lahris (5,400), Bangulzais (9,000), Kūrds (3,100), Shāhwānis (6,300), Muhammad Shāhis (2,800), Raisānis (2,400), and Sarparras (900), all of whom are cultivators and flock-owners. In this category must also be included the numerous Lāngav cultivators of Mungachar (17,000). All the Muhammadans are of the Sunni sect. A few Hindu traders are scattered here and there. Most of the wealthier men possess servile dependants. Artisans' work is done by Loris. The prevailing language is Brāhui; but the Lāngavs, some of the Bangulzais, and a few other clans speak Baluchi, and the Dehwārs a corrupted form of Persian.

Agriculture.

Cultivation is carried on in the centre of the valleys, which possess flat plains of a reddish clay soil, highly fertile when irrigated. This is the best soil and is known as *matt, matmāl,* or *hanaina.* Dark loam is known as *siyāhzamīn.* The greater part of the cultivable area is 'dry' crop (*khushkāba*). Owing to the scanty rainfall, it seldom produces a full out-turn oftener than once in four or five years. The principal 'dry' crop areas are Narmuk, Gwanden, the Bhalla Dasht or Dasht-i-bedaulat,

Irrigation and crops.

Kābo, Kūak, Khad, the Chhappar valley, and Gurgīna. Kalāt, Mungachar, Mastung, and Johān are the best irrigated areas. Irrigation is derived from underground water-channels (*kārez*), which number 247, from springs, and from streams. Many of the *kārez* are dry at present (1906). Fine springs occur at Kahnak in Mastung, at Kalāt, Dudrān near Chhappar, and Iskalku; and the Sarawān and Shīrīnāb rivers afford a small amount of irrigation. The principal crop is wheat, the flour of which is the best in Baluchistān. In 'wet' crop areas lucerne, tobacco, and melons are produced in large quantities. Johān tobacco is famous. The cultivation of onions and potatoes is increasing. Fine orchards are to be seen at Mastung and Kalāt; and in the former place, where mulberries abound, experiments are being made in the introduction of sericulture.

The sheep are of the fat-tailed variety, and goats and camels are numerous. The best of the latter are to be found in Mungachar. Fine horses are bred, the principal breeders being the Shāhwānis, Garrāni Bangulzais, Muhammad Shāhis, and some Lāngavs. The number of branded mares is 179, and 12 Government stallions are at stud in summer. Mungachar donkeys are of large size. The bullocks are short and thick-set. Camels, horses, and cattle.

The chief forest tract is the Harboi range, which is well covered with juniper. Pistachio forests also occur here and there. Tribal rights exist in most of the forests, and portions are occasionally reserved for fodder. No systematic reservation is attempted by the State. Great care of pistachio trees is taken by the people when the fruit is ripening. Coal is worked in the Sor range, and traces of the same mineral have been found near Mastung. Ferrous sulphate exists in the Melabi mountain. Forests and minerals.

The wool of sheep and goats, of which there is a large production in the country, is utilized in the manufacture of felts (*thappur*), rugs in the *darī* stitch (*kont* and *shifī*), saddle-bags (*khurjīn*), and overcoats (*zor* and *shāl*). The best rugs are manufactured by the Badduzai clan of the Bangulzais. All women do excellent needlework. Embroidered shoes and sandals, which are made at Kalāt and Mastung, are popular. Arts and manufactures.

The chief trading centres are Mastung and Kalāt. The exports consist chiefly of wool, *ghī*, wheat, tobacco, melons, carbonate of soda, sheep, and medicinal drugs; and the imports of cotton cloth, salt, iron, sugar, dates, and green tea. Caravans carry tobacco, wheat, and cloth to Panjgūr in Makrān, and return laden with dates. Commerce.

The Mushkāf-Bolān section of the North-Western Railway touches the country, and the Quetta-Nushki line traverses its northern end. A metalled road, 88½ miles long, built in 1897 and since slightly improved, at a total cost of 3¼ lakhs of rupees, Railways and roads.

runs from Quetta to Kalāt. Communications from north to south are easy. From west to east the tracks follow two main lines : from Kardgāp through the Mastung valley and over the Nishpa Pass to the Bolān, and through Mungachar and Johān to Narmuk and to Bībī Nāni in the Bolān Pass. Communications with the Mastung valley are being improved by the construction of tracks over several of the passes.

Famine.　　The country is liable to frequent scarcity, but owing to the number of *kārez* it is the best-protected part of the States. The nomadic habits of the people afford a safeguard against famine ; and, even in years when rainfall is insufficient for 'dry' crop cultivation, they manage to subsist on the produce of their flocks, supplemented by a small quantity of grain.

Subdivisions, staff, criminal and civil justice.　　For purposes of administration the people, rather than the area, may be divided into two sections : namely, those subject to the direct jurisdiction of the Khān of Kalāt, and those belonging to tribal groups. The principal groups constituting each section have been named above. The areas subject to the Khān are divided into the two *niābats* of Mastung and Kalāt. The Mastung *niābat* forms the charge of a *mustaufi*, who is assisted by a *naib* and a *jā-nashīn*. Kalāt is in charge of a *naib*. The Brāhui tribesmen are subject to the control of their chiefs, who in their turn are supervised by the Political Agent through the Native Assistant for the Sarawān country and the Political Adviser to the Khān. For this purpose *thānadārs*, recruited from the Brāhuis, are posted at Alu, Mastung, and Mungachar. In the Khān's *niābats* the various officials deal with both civil and criminal cases, subject to the supervision of the Political Adviser to the Khān. Cases among the tribesmen, or cases occurring between subjects of the Khān and the tribesmen, are disposed of by the Political Agent or his staff, and are generally referred to *jirgas*. Cases for the possession of land or of inheritance are sometimes determined by local *kāzis* according to Muhammadan law.

Land revenue.　　Mastung and Kalāt-i-Nichāra, i.e. Kalāt and the neighbourhood, are mentioned in the *Ain-i-Akbarī* as paying revenue in kind and furnishing militia to Akbar. The only part of the country which has been surveyed is Kahnak, where, owing to disputes between the Rustamzai clan of the Raisāni tribe and the chief section, a record-of-rights was made in 1899. The land is vested in a body of cultivating proprietors, who either pay revenue or hold revenue-free. The rate of revenue varies from one-fourth to one-tenth of the produce, and is generally taken either by appraisement or by an actual share. Of the areas subject to the Khān, the revenue of Johān with Gazg is leased for an annual payment in kind, and the same system is followed in other scattered tracts. In the Kalāt *niābat*, revenue is paid by the cultivators either in kind or in personal service as horsemen, footmen, labourers, and messengers. In Mastung the land revenue is recovered both in kind and at a fixed rate in cash and kind (*zarri* and *kalang*). In the case of many of the *kārez* in the Mastung and Kalāt *niābats*, the State, to avoid the trouble of collecting the produce revenue at each harvest, has acquired a proportion of the land and water supplied by a *kārez* in perpetuity and converted them into crown property, leaving the remainder of the land and water free of assessment. In 1903 the revenue of each *niābat* was as follows : Mastung, Rs. 92,800 ; Kalāt, Rs. 32,700; Johān with Gazg, Rs. 1,200 ; total, Rs. 1,26,700.

KALĀT TOWN is the head-quarters of the Khān's military forces, and a regiment of cavalry, ninety-five sabres strong, is stationed at Mastung. Tribal levies, thirty-two in number, are posted at Mastung, Alu, and Mungachar. Irregular levies, to the number of eighty-six, maintained by the Khān for the collection of revenue and keeping the peace in his own *niābats*, are stationed at Kalāt. There is a small jail at Mastung and a lock-up at Alu. *Army, levies, and jails.*

During the second Afghān War, the Sarawān chiefs rendered good service in guarding communications and providing supplies, in recognition of which the British Government granted personal allowances to some of them. These payments have since been continued, to assist the *sardārs* in maintaining their prestige and in keeping order among their tribesmen, and amount to Rs. 22,800 per annum.

Education is neglected. A few persons of the better class keep *mullās* to teach their sons, and a school, which promises to be well attended, is about to be opened at Mastung. Two dispensaries are maintained, one by the British Government and the other by the Kalāt State. The total number of patients in 1903 was 8,919, and the total cost Rs. 5,300. Inoculation is practised by Saiyids, who generally get fees at the rate of eight annas for a boy and four annas for a girl. *Education and medical.*

Kalāt Town.—Capital of the Kalāt State in Baluchistān, situated in 29° 2′ N. and 66° 35′ E., 88½ miles from Quetta on the south of the Sarawān division. It is known to the natives as Kalāt-i-Baloch and Kalāt-i-Sewa ; the former to distinguish it from Kalāt-i-Ghilzai in Afghānistān, and the latter after its legendary founder. The population does not exceed

2,000 persons (1901). The inhabitants are chiefly the Khān's troops, numbering 491, and his retainers, with a few Hindu traders. The town occupies a spur of the Shāh-i-Mardān hill on the west of the Kalāt valley. A wall surrounds it, with bastions at intervals. Its three approaches on the north, south, and east are known respectively as the Mastungi, Gilkand, and Dildār gates. Three suburbs lie close by. Commanding the town is the *miri* or citadel, an imposing structure in which the Khān resides. Kalāt fell into the hands of the Mīrwāris about the fifteenth century, since which time the place has remained the capital of the Ahmadzai Khāns. In 1758 it withstood three assaults by Ahmad Shāh Durrani, and in 1839 was taken by the British under General Willshire. A year later it surrendered to the Sarawān insurgents. Below the citadel lies a Hindu temple of Kālī, probably of pre-Muhammadan date. The marble image of the goddess, holding the emblem of plenty, stands in front of two lights which are perpetually burning. The trade of the town is chiefly retail business. Taxes on trade are collected by a system of contracts. Police functions are carried out by an official known as *mīr shab*, assisted by watchmen (*kotwāls*).

Boundaries, configuration, and hill and river systems.

Kachhi.—A division of the Kalāt State, Baluchistān, lying between 27° 53′ and 29° 35′ N. and 67° 11′ and 68° 28′ E. It consists of a flat triangular plain, 5,310 square miles in area, with its base on the Upper Sind Frontier District of Sind, and is enclosed by the Marri and Bugti hills on the east, and by the Kīrthar and Central Brāhui ranges of the Jhalawān country on the west. On the north-east side of its apex lies the British *tahsīl* of Sibi. The only hills, other than the skirts of the surrounding mountains, consist of the low range called Bannh, separating Dādhar on the north from the Bolān lands on the south. The principal rivers are the NĀRI, the Bolān, the Sukleji, and the MŪLA. Among the hill-torrents are the Dhoriri, formed by the junction of the Sain and Karu from the Jhalawān country, the Lahri, and the Chhatr. On entering Kachhi, all these rivers are dissipated into numberless natural channels, spreading over the great alluvial stretches of which the country is composed.

Geology, botany, and fauna.

The geological structure of the district is uniform, consisting of a level bed of clay burnt by the rays of the sun and probably of great depth. The outskirts of the surrounding hills are of the Siwālik formation. Except along the foot of the hills, the general aspect of the country is desolate and bare. The Flora is thorny and scant, consisting of a stunted scrub.

Among the trees occur *Prosopis spicigera, Salvadora oleoides, Capparis aphylla,* and two kinds of *Tamarix.* Common plants are *Calotropis procera* and many saltworts, such as *Haloxylon salicornicum.* Wild animals are scarce; a few ravine and other small deer occur, and flocks of sand-grouse visit the cultivated areas in winter.

Situated in close proximity to Sind, Kachhi is one of the hottest areas in India. Scorching winds blow in the summer, and at times the deadly simoom (*luk*) prevails. Mosquitoes are so numerous that, at Gājān, a special portion of the crop has been assigned to a saint for his protection against them. From November to February the climatic conditions are pleasant, the air being crisp and cool. The annual rainfall averages about 3 inches and usually occurs in July and August.

Climate, tempera- ture, and rainfall.

The history of Kachhi is intimately connected with that of Sind. In the seventh century Rai Chach took its capital Kandābīl, probably Gandāva. To the Arabs the country was known as Nūdha or Būdha, and Kandābīl was despoiled by them on several occasions. It afterwards passed into the hands of the Sūmras and Sammas of Sind. The fifteenth century saw the arrival of the Baloch and the conflicts between their two leaders, Mīr Chākar, the Rind, and Gwahrām Lāshāri. The Arghūns next took possession, and from them the country passed to the Mughals, and on the decline of the latter to the Kalhoras. In 1740 Nādir Shāh handed it over to the Brāhuis in compensation for the death of Mīr Abdullah, the Ahmadzai Khān of Kalāt, at the hands of the Kalhoras in the fierce battle of Jāndrīhar near Sanni. From 1839 to 1842, during the first Afghān War, Kachhi was held and administered by the British on their lines of communication, and was the scene of much raiding and of two fights with the insurgent Brāhuis in 1840. After the war General John Jacob's cavalry was employed in checking the raiding propensities of the Kachhi tribesmen, especially the Jakrānis, who were subsequently removed to Sind. In the time of Mīr Khudādād Khān of Kalāt it was long a scene of anarchy and raiding, and at Bhāg in 1893 this ruler committed the crime in consequence of which he subsequently abdicated.

History and archaeo- logy.

Buddhist remains have been discovered at Chhalgari and Tambu, and many of the mounds scattered through the country would probably repay excavation.

The number of villages is 606. The population (1901) is 82,909, the majority being Jats. Among important Baloch tribes are the Rinds, Magassis, and Lāshāris; and among minor

The people, their tribes and occupa- tions.

tribes, Buledis, Dombkis, Kaheris, and Umrānis. Roughly speaking the Magassis and Rinds occupy the west, and the Dombkis, Kaheris, and Umrānis the east; the Jats are found everywhere as cultivators. A few Brāhuis, such as the Raisānis and Garrāni Bangulzais, have permanently settled in the north of the country, and in the cold weather it is visited by many other Brāhuis from the highlands. The occupation of nearly all the people is agriculture. Hindu traders are found in all important villages; the lower castes include potters, sweepers, blacksmiths, and weavers. The most common language is Sindī, but Western Punjābi and Baluchi are also spoken. Except the Hindu traders, all the people are Sunni Musalmāns. A sect called Taib ('penitents') has made some progress since 1890.

Agriculture.
It is usual to speak of Kachhi as a desert, but this is a mistake. The soil is extremely fertile wherever it can be irrigated. Its quality depends on the admixture of sand. The best is a light loam mixed with a moderate proportion of sand (*matt*). Except a fringe of 'wet' crop area on the west, most of the land entirely depends for cultivation on floods brought down by the rivers from the surrounding hills, the water of which is raised to the surface by a system of large dams constructed in the beds of the rivers by the co-operation of the cultivators. A description of this interesting system will be found in the paragraphs on agriculture in the article on BALUCHISTĀN. The floods generally occur in July and August, but occasionally also in spring. Three crops are harvested during the year: *sānwanri*, *sarav*, and *arhāri*. The first is the principal crop, and is sown in July and August and reaped in the autumn. It consists of *jowār* with a little *mūng*, *moth*, and *bājra*. The second, or spring crop, comprises wheat, barley, mustard, and rape; the third, *jowār* for fodder, cotton, and water-melons. Kachhi *jowār* is renowned for its excellence, and is usually cultivated on a soil known as *khauri*. Indigo is grown in Dādhar. The cultivation of the *sarav* crop is uncertain, depending on late floods in August. Dādhar, Sanni, Shorān, Gājān, Kunāra, part of Gandāva, and Jhal are the only places where irrigation from permanent sources exists.

Cattle, horses, sheep, &c.
Bullocks from Nāri in Kachhi are famous for their shape and strength, and many are purchased by dealers from the Punjab. Camels are bred in some numbers. The breed of horses is excellent. Branded mares number 604, and one stallion was located in the country in April, 1904. The best breeders are the Magassis, Dombkis, and Rinds. The indi-

genous sheep do not possess fat tails, but many of the fat-tailed variety, known as Khorāsāni, are brought from the highlands in winter. Of goats the *barbari* breed is most prized.

No 'reserved' forests exist, but protective measures are Forests adopted by the tribal chiefs. The western side of the country and minerals. contains some well-wooded tracts. A sulphur mine at Sanni was worked in pre-British days by the Amīrs of Afghānistān. Ferrous sulphate (*zāgh*) is found in the mountains near Kotra and Sanni. Earth-salt is manufactured by the lixiviation of salt-bearing earth at Gājān and Shorān. Saltpetre is produced in small quantities, and the manufacture of carbonate of soda (*khar*) from the numerous saltworts is increasing.

The principal industry is the weaving of coarse cotton cloth. Manufac-Double coloured cotton sheets (*khes*) of good quality are tures. produced here and there, while at Lahri and a few other places a fine kind of embroidered leather-work is manufactured. Country rifles, swords, and saddles are made at Bhāg and Dādhar.

Most of the trading class come from Shikārpur in Sind. Commerce. The centres of trade are Dādhar, Lahri, Hāji, Bhāg, Shorān, Gājān, Kotra, Gandāva, and Jhal. Piece-goods, rice, sugar, and country carts are imported from Sind; dates, *ghī*, wool, and medicinal drugs from the highlands for re-export. Exports to the highlands include cotton cloth, mustard oil, salt, and silk; the articles supplied to Sind consist chiefly of carbonate of soda, grain, and oilseeds. The North-Western Railway passes through the centre of Kachhi. No metalled roads have been made, but the country is easily traversed in all directions except after heavy floods.

The principal routes run from Jacobābād to Sibi via Lahri Communi-on the east; through Shori and Bhāg in the centre; and via cations. Gandāva and Shorān to Dādhar on the west. The route through the Mūla Pass from the Jhalawān country debouches at Gandāva.

The insignificant rainfall, the dependence of the country on Famine. flood-irrigation, and the absence of proper means of distributing the flood-water render Kachhi extremely liable to scarcity and even to famine. Under existing conditions enormous quantities of water run to waste in the Nāri in ordinary years, and the introduction of a good irrigation and distribution scheme would doubtless afford a large measure of protection. The proximity of Sind and the free migratory habits of the population have hitherto prevented the necessity of actual famine relief. Advances amounting to about Rs. 29,000 were made to the Khān of Kalāt's cultivators in 1900, when the drought, which had begun in 1897, culminated. They were recovered at the succeeding harvests.

<p>

Subdivisions and staff.

For purposes of administration, Kachhi is divided into two parts : areas subject to the jurisdiction of the Khān of Kalāt, and areas under tribal chiefs. Within the areas subject to the Khān, however, tribal units are to be found which occupy a position of practical independence. The political control of the country east of the railway, i.e. the whole of the Lahri *niābat*, is vested in the Political Agent of the Sibi District, and of the remainder in the Political Agent, Kalāt. The area under the Kalāt State is divided into five *niābats*: Dādhar; Bhāg ; Lahri, which includes the area occupied by the Dombki, Kaheri, and Umrāni tribes ; Gandāva ; and Nasīrābād. The head-quarters station of each *niābat* is located at a village of the same name, except Nasīrābād, of which the head-quarters are situated at Mīrpur Bībīwāri. Dādhar, Bhāg combined with Lahri, and Gandāva with Nasīrābād each forms the charge of a *mustaufī*, who is generally assisted by local officials known as *naib* and *jā-nashīn*. Dādhar, however, possesses neither a *naib* nor a *jā-nashīn*, and Gandāva has no *jā-nashīn*. The principal areas subject to tribal control are Jhal, inhabited by the Magassis ; and Shorān, held by the Rinds. In Lahri, the Dombkis, Kaheris, and Umrānis have acquired a large measure of independence. In the *niābats*, criminal, civil, and revenue cases are decided by the local officials ; in tribal areas, petty cases are dealt with by the chief, and important or intertribal cases are referred to *jirgas* or local *kāzīs*, who exercise much influence, under the orders of the Political Agents. In the numerous *jāgīrs* within the Khan's *niābats*, jurisdiction in all petty matters is exercised by the *jāgīrdārs*. The most common offences are cattle-lifting and theft. Cattle are frequently stolen from Sind and sent to the Jhalawān country. Much use is made of trackers in the detection of such crimes, and some of these men are very skilful. They are paid by results.

Land revenue.

The land revenue system presents an interesting survival of ancient native methods. The Khān collects revenue in his *niābats*, and elsewhere it is taken by tribal chiefs and *jāgīrdārs*. It consists everywhere of a fixed share of the gross produce, varying from one-third to one-tenth, but generally one-third or one-fifth. The additional cesses (*rasūm*), however, raise the amount paid to one-half. Irrigated lands sometimes pay a fixed cash assessment (*kalang*). Large *jāgīrs*, originally granted for feudal service, are held by the Sarawān tribesmen in Bālā

Nāri and the Bolān lands, and by the Jhalawān tribes round Gandāva and at Chhatr-Phuleji. The Dombki headman holds one in Lahri. Generally the proprietary right in all areas is held by the local cultivating class, but in the Baloch areas of Jhal and Shorān it has been transferred in many cases to the chiefs.

Besides the land revenue, contracts are given in the *niābats* for octroi, excise, and the collection of other minor taxes, the proceeds being included in the total revenue. The amount of land revenue proper varies with the extent and time of the floods in the rivers. Thus, in 1902 the Khān's aggregate revenue from all his *niābats* amounted to about 1 lakh, and in 1903 to more than 2½ lakhs; but in the latter year a new system of administration had been introduced. The details of the latter sum are as follows : Dādhar, Rs. 49,200 ; Bhāg, Rs. 32,500; Lahri, Rs. 58,100 ; Gandāva, Rs. 55,400 ; Nasīr-ābād, Rs. 56,600.

Tribal levies, paid by the British Government and number- Army, ing fifty, are stationed at Dandor in Bālā Nāri, Lahri, Phuleji, levies, jails, and along the railway. Detachments, consisting of eighty-five education, of the Khān's infantry and twelve artillerymen, are located at and medical. Dādhar, Nasīrābād, and Bhāg; but their numbers vary from time to time. The number of the Khān's irregular levies is generally ninety-one. A tribal *thāna* of five men is posted at Gandāva. Security is provided by the enlistment of *kotwāls*, who are paid either by the inhabitants or from the Khān's revenues. Tribal chiefs maintain retainers and dependants, who are employed on revenue duties and in securing the general peace. The same system is followed in the Khān's *niābats* by the local *naibs*, who distribute their friends and followers throughout the country at the expense of the culti-vators. The Rind and Magassi chiefs receive allowances from the Khān of Kalāt of Rs. 300 a month each. A jail is now in course of erection at Dādhar; criminals have hitherto generally been kept in the stocks. The country has no schools or dis-pensaries. Inoculation takes the place of vaccination, being performed by Saiyids, Pīrs, Shehs (the local name for Shaikhs), and Abābakis from the highlands.

Jhalawān.—A highland division of the Kalāt State, Boun-Baluchistān, comprising the country to the south of Kalāt daries, con-figuration, as distinguished from Sarawān, the country to the north of and hill that place, and lying between 25° 28′ and 29° 21′ N. and and river systems. 65° 11′ and 67° 27′ E., with an area of 21,128 square miles. It is bounded on the north by the Sarawān country; on the south by the Las Bela State ; on the east by Kachhi and

Sind ; and on the west by Khārān and Makrān. The boundary
between the Jhalawān country and Sind was settled in 1853-4
and demarcated in 1861-2. Elsewhere it is still undetermined.
An imaginary line drawn east and west through Bāghwāna
divides the country into two natural divisions. To the north
the general conditions are those of the upper highlands, and
to the south those of the lower highlands of Baluchistān. The
country has a gradual slope to the south, with valleys of considerable width lying among lofty mountain ranges. Among
the more important valleys are Sūrāb with Gidar, Bāghwāna,
Zahri, Khuzdār with Fīrozābād, Wad, Nāl, Sārūna, Jau, and
the valley of the Mashkai river. The mountains comprise the
southern portion of the CENTRAL BRĀHUI range, including
the Harboi hills and the greater part of the KĪRTHAR and PAB
ranges. On the west the Garr hills and their continuation
southward separate the country from Khārān and Makrān ; in
its centre lie a number of more or less detached mountains, the
chief of which are Dobānzil (7,347 feet), Hushtir (7,260 feet),
Shāshān (7,513 feet), and Drā Khel (8,102 feet). The rivers
include the HINGOL, the largest river in Baluchistān, with its
tributaries the Mashkai and the Arra ; the MŪLA, the HAB,
and a portion of the PORĀLI. Among the less important
streams may be mentioned the Karkh or Karu and the Sain
rivers, which debouch into the Kachhi plain near Jhal, and
the Kolāchi or Gāj, which traverses the centre of the country.
None of these rivers possesses a continuous flow of water
throughout its course.

Geology.   In the north of the country nummulitic limestone is met
with. Farther south red and white compact limestone (upper
cretaceous) is very extensively distributed. Beds containing
chert are of frequent occurrence. Igneous rocks occur near Nāl,
and on the east are the Kīrthar, Nāri, and Gāj geological groups.

Botany.   Vegetation is scanty except in the Harboi hills, on the north,
where juniper and wild almond grow in abundance. Else-
where olive and pistachio occur. In the south the little tree
growth includes *Capparis aphylla, Prosopis spicigera*, two kinds
of *Acacia*, and *Tamarix articulata*. The northern hill-sides
are thickly covered with a scrub jungle of *Artemisia* and
Haloxylon Griffithii. Tulips, irises, and other bulbous plants
appear in the spring. The grasses are of the orders *Bromus,
Poa*, and *Hordeum*. Dwarf-palm (*Nannorhops Ritchieana*)
grows in profusion in the lower hills. Pomegranates are the
commonest trees in the gardens, but apricots, mulberries, and
dates are also found.

Sind ibex and mountain sheep are the most common game, Fauna. but their numbers are decreasing. Leopards and black bears are occasionally killed. Wild hog are met with in the Mashkai valley. Ravine-deer are common. A few grey and black partridges are to be found in Lower Jhalawān; *chikor* are numerous in the higher parts, and *sīsī* almost everywhere.

Upper Jhalawān possesses a climate resembling that of Quetta, moderate in summer and cold in winter, with well-marked seasons. The lower parts are pleasant in winter, but subject to intense heat in summer. At this time fever is very prevalent in places south of the Harboi range. Earthquakes frequently occur in the neighbourhood of Wad and Mashkai. The rainfall is scanty, and is received in the upper highlands in winter and in the lower parts in spring or summer. *Climate, temperature, and rainfall.*

The country passed in the seventh century from the Rai dynasty of Sind to the Arabs, by whom it was known as Turān. Its capital was Khuzdār, which place was also the head-quarters of the Arab general commanding the Indian frontier. Kaikānān, probably the modern Nāl, was another place of importance. The Ghaznivids and Ghorids next held the country, and were followed by the Mongols, the advent of Chingiz Khān being still commemorated by the Chingiz Khān rock between Nichāra and Pandrān. With the rise of the Sūmra and Samma dynasties in Sind, the Jat aborigines of the country appear to have gained the ascendancy, but in the middle of the fifteenth century they were ousted by the Mīr-wāris. Beginning from Nighār near Sūrāb, these founders of the Brāhui kingdom gradually extended their dominion over all the Jhalawān hills. For many years the country remained firmly attached to the Khāns of Kalāt; but the struggles which took place during Mīr Khudādād Khān's reign involved the Jhalawān tribesmen also and resulted in the strangling of their leader, Tāj Muhammad, Zarakzai. In 1869 Jām Mīr Khān of Las Bela, who had caused the people of Jhalawān to rebel under Nūr-ud-dīn Mengal, received a severe defeat in a battle near Khuzdār, when he lost seven guns. Owing to its remoteness from Quetta, the Jhalawān country did not come so quickly and completely under control after the British occupation as the Sarawān country; and an outbreak which began in 1893 under the leadership of Gauhar Khān, the Zahri chief, simmered till 1895, when it was put down by the Kalāt State troops at the fight of Garmāp, in which both Gauhar Khān and his son lost their lives. *History and archaeology.*

The country is comparatively rich in archaeological remains.

They include many *gabrbands* or embankments of the fire-worshippers; a curious vaulted burial chamber cut in the slope of the hill near Pandrān; and several tombs which indicate a system of superterrene burial. Interesting earthen vessels, and stones bearing Kufic inscriptions, have been excavated from the numerous mounds in the country.

The people, their tribes and occupations.

Jhalawān contains no large towns and only 299 permanent villages. KHUZDĀR is the head-quarters station. Most of the people live in blanket tents or mat huts. The inhabitants, the majority of whom are Brāhuis with here and there a few Baloch, Jats, and Loris, number 224,073 (1901), or about eleven persons to the square mile. They include the direct subjects of the Khān, such as Kūrds, Nighāris, Gazgis, and Nakībs, who cultivate lands in the Khān's *niābats*; and tribal units. The principal tribes are the Zahri (49,000), the Mengal (69,000), the Muhammad Hasni (53,000), and the Bīzanjau (14,000). Among minor tribal groups may be mentioned the Sājdi, Rodeni, Rekizai, Gurgnāri, Sumalāri, Kambrāni, Mīrwāri, and Kalandarāni. The leading chief of the Jhalawān tribes belongs to the Zarakzai clan of the Zahri tribe. A few Hindus carry on the trade of the country. Most of the people speak Brāhui; a few speak Sindī; the remainder, especially on the south-west, Baluchi. The majority of the people are Sunni Muhammadans, but some are Zikris, especially the Sājdis. Agriculture and flock-owning are the only occupations. Every year in September, a large migration of nomads takes place to Kachhi and Sind, where they engage in harvesting and return to the highlands in spring.

General agricultural conditions. Irrigation.

Cultivation is confined to the valleys and the flats beside the river courses. Most of the cultivated tracts consist of 'dry' crop areas, dependent on flood-water which is held up by embankments. In comparison with the Sarawān country irrigation is scarce. It is obtained from springs, from *kārez*, which number only thirty-five, and from channels cut from the rivers. Most of the springs and *kārez* occur in Upper Jhalawān. Tracts irrigated by river water include Zahri, part of Gidar, Khuzdār and Zidi, Karkh and Chakku, the valleys of the Mūla and Kolāchi rivers, and Mashkai. Well-irrigation is unknown. The soil has a considerable mixture of sand, and is but moderately fertile. 'Dry' crop areas produce better crops than 'wet' areas, unless the latter are highly manured.

Principal crops.

The spring harvest is the most important, consisting chiefly of wheat. On the south-west, however, wheat suffers from the damp caused by the sea-breezes, and its place is taken by barley. Rice is grown along the banks of the Mūla and Kolāchi rivers, and, with *jowār*, forms the chief autumn harvest. Dates are grown in Mashkai. Cultivation is gradually extending, but the people prefer flock-owning to cultivation, and progress is slow. Jhalawān is in fact a vast grazing tract.

The bullocks are hardy but small, and a good many are bred Cattle, in the lower tracts. Sheep and goats are found in vast numbers. horses, The Khorāsāni variety of sheep is preferred to the indigenous sheep, &c. kind, owing to its larger tail. Most of the camels are transport animals, and camel-breeding is almost entirely confined to the Pab range. A few horses are kept in the north, but they are not so numerous as in the Sarawān country.

Lead-smelting was carried on in former days at Sekrān near Minerals. Khuzdār, and Masson mentions the employment of 200 men in 1840, but the industry has now been abandoned. Little is known of the other minerals of the country. Ferrous sulphate (melanterite), known locally as *zāgh* or *khāghal*, has been found in the Ledav river and near Zahri. A soft ferruginous lithomarge, known as *mak*, is used as a mordant in dyeing.

The manufacture of coarse woollen rugs in the *darī* stitch, Arts, and of felts, ropes, and bags, is general; good pile carpets are manufactures, and woven for private use by the Bādīnzai Kalandarānis of Tūtak commerce. and at a few other places. Nīchāra needlework is famous locally. There is a large export to Sind of matting and materials for mat-making, and many of the people entirely depend on this source of livelihood. The chief centres of trade are Sūrāb, Khuzdār, Nāl, Wad, and Mashkai; but trading is much hindered by the levy of transit-dues, by both the Kalāt State and local chiefs. *Ghī*, wool, live sheep, and materials for mats are the principal exports; coarse cloth, sugar, mustard oil, and *jowār* are imported.

Railways and metalled roads do not exist. Travellers Communifollow camel-tracks, the most important of which are the cations. Kalāt-Bela route, known as the Kohan-wāt, via Khuzdār, Wad, and the Bārān Lak; the Kalāt-Panjgūr route via Sūrāb; and the Kachhi-Makrān route via the Mūla Pass, Khuzdār, Nāl, and Mashkai. An unmetalled road is now (1906) under construction between Kalāt and Wad.

Drought is frequent, owing to the shortness of the rainfall, but Famine. the proximity of Sind enables the inhabitants to find a ready means of support at such times. During a drought of exceptional severity, which began in 1897 and culminated in 1901, Brāhuis were known in several instances to have taken their daughters of marriageable age to Sind, where the high brideprices obtained for them enabled the parents to tide over the bad times. Cases have also been known in which servile dependants were exchanged for a maund of dates.

Administration, staff, &c. Since 1903 an officer, known as the Native Assistant for the Jhalawān country, has been posted to Khuzdār by the Khān of Kalāt under the supervision of the Political Agent. He is supported by twenty levies, and decides petty intertribal and other cases with the assistance of *jirgas*. For administrative purposes, the country consists of areas subject to the Khān of Kalāt and of tribal areas. The former include the *niābats* of Sūrāb and Khuzdār, each of which is in charge of a *naib*. The former has a *jā-nashīn*, or assistant, stationed at Mashkai; and the latter have three *jā-nashīns*, stationed at Karkh or Karu, Zidi, and Bāghwāna. The Khān's interests in Zahri are supervised by a *daroga*. His rights at Gazg are leased to a farmer with those of Johān in the Sarawān country. In former times the Khān's *naibs* exercised a general control and communicated the Khān's orders to the tribal chiefs; but the latter are now largely controlled by the Political Agent through the Native Assistant in the Jhalawān country. They decide cases occurring among their tribesmen according to tribal custom. In civil suits, a custom has been established of taking one-fourth of the value of the property decreed.

Revenue. Land revenue, in the case of the subjects of the Khān, is always taken in produce, the rates generally varying from one-fourth to one-eighth. Cesses, known as *rasūm* or *lawāzimāt*, are also levied, by which the State share is largely increased. Transit-dues, and fines known as *bādi hawai*, constitute the other sources of revenue in the *niābats*. Contrary to the custom in the Sarawān country, the Jhalawān chiefs exact *mālia* from their tribesmen, generally in the shape of one sheep per household annually. Sheep are also taken on marriages and other festivals in a chief's household and on the occurrence of deaths. These payments are known as *bijār* and *purs*. Some of the chiefs also levy octroi and transit-dues. The value of the total revenue from the Khān's *niābats* varies with the agricultural conditions of the year. In 1903–4 the approximate amounts received were as follows : Sūrāb with Mashkai, Rs. 15,500; Khuzdār, Rs. 14,200; Zahri, Rs. 1,300; total, Rs. 31,000.

Levies. In 1894, owing to the unsatisfactory state of affairs in Jhalawān, the Khān sanctioned payments aggregating about Rs. 40,000 per annum to the chiefs of the principal tribes, in return for which they were made responsible for the peace of their respective areas. This sum included the allowances of the Rind and Magassi chiefs in Kachhi. A sum of Rs. 3,600 is also contributed by the British Government. At this time tribal levy posts of ten men each were also instituted at Zahri and Sārūna. A post of ten men has since been stationed at Sūrāb, besides the Native Assistant's levies at Khuzdār. The *naib* of Khuzdār is assisted by forty-five levies for revenue and police purposes, and the *naib* of Sūrāb by twenty-five men ; but these numbers are increased or reduced as occasion requires. The *naibs* and stronger chiefs generally have stocks in their forts or houses in which offenders are placed.

A few of the chiefs employ Afghān *mullās* for teaching their sons; otherwise education is unknown. The people are very superstitious, and have a firm belief in the influence of evil spirits, to whom diseases are generally attributed. A few resort to the dispensary at Kalāt for medical treatment. They are well versed in the use, as remedies, of the medicinal drugs which the country produces in large quantities. The sick are frequently branded ; for fevers the usual remedy is to wrap the patient in the fresh skin of a sheep or goat. Inoculation by Saiyids is general, except in the case of the Zikris. *Education and medical.*

Khuzdār.—The principal place in the Jhalawān division of the Kalāt State, Baluchistān, and the head-quarters of the Native Assistant and of the Khān of Kalāt's *naib*, situated in 27° 48′ N. and 66° 37′ E. It is known to the Sindīs as Kohiār, and is a long narrow valley at the upper end of which a fort was constructed in 1870. Khuzdār owes its importance to its central position at the point of convergence of roads from Kalāt on the north, Karāchi and Bela on the south, Kachhi on the east, and Makrān and Khārān on the west. It is unhealthy in summer. The garrison consists of seven artillerymen with one gun and forty-five irregular levies. The Native Assistant has a small escort of twenty levies. The *niābat* of Khuzdār includes land in Bāghwāna, Zidi, the valley of the Kolāchi river, Karkh and Chakku.

Khārān.—A *quasi*-independent tribal area of the Kalāt State, Baluchistān, lying between 26° 52′ and 29° 13′ N. and 62° 49′ and 66° 4′ E., with an area of 14,210 square miles. It consists of a wide plain, irregularly quadrilateral in shape, and varying in elevation from 2,500 feet on the north-east to 1,600 feet on the west. It is bounded on the north by the Rās Koh hills ; on the south by the Siāhān range ; on the east by the Garr hills of the Jhalawān country ; while on the west the boundary runs with Persia. The country is generally regarded *Boundaries, configuration, and river systems.*

as entirely desert; in reality, however, considerable tracts of cultivated land are situated at the foot of the hills and along the courses of the Baddo and Māshkel rivers. Most of the remainder of the country is covered by immense stretches of sand. The hydrography of the plain is peculiar. Torrents drain into it from the surrounding mountains, but find no outlet to the sea. Besides the Māshkel and Baddo, the only streams of importance are the Garruk or Sarāp and the Korakān.

Geology, botany, and fauna.
The only part of Khārān that has been geologically examined is the Rās Koh range, the mass of which may be divided into three zones, the northern consisting mainly of intrusive rocks, the central of shales, and the southern of tall limestone ridges. The plain is covered in parts with alluvial deposit and elsewhere with sand. The botany of the country has never been studied. Trees are scarce, but the ravines contain quantities of tamarisk, of *Haloxylon ammodendron*, and in years of good rainfall many grasses. Among the latter may be mentioned *magher* (*Rumex vesicarius*), the seed of which is eaten as a famine food and is also exported. Another famine food consists of *kulkusht* (*Citrullus Colocynthis*), the seeds of which are made into bread. The surrounding hills produce asafoetida. Sind ibex and mountain sheep inhabit the hills, and ravine-deer their skirts. Herds of wild asses are found in the neighbourhood of the Māshkel river. Snakes are numerous.

Climate, temperature, and rainfall.
The climate is dry but healthy. Severe dust-storms are experienced throughout the year, being specially trying from June to September. The heat in summer is great, but the nights are always cool. The winter is cold. Most of the small amount of rain which falls is received between January and March.

History and archaeology.
Little is known of the history of the country previous to the end of the seventeenth century, when Ibrāhīm Khān, the Nausherwāni chief of Khārān, served the Ghilzai dynasty of Kandahār, except that it appears to have formed part of the Persian province of Kirmān. The Nausherwāni chiefs, round whom local history centres, claim descent from the Kiānian Maliks, and have always been a race of strong-willed, bold, and adventurous men, taking full advantage of their desert-protected country for organizing raiding expeditions against their neighbours, and professing a fitful allegiance to Persia, to Kalāt, and to Afghanistān in turn. The most famous were Purdil Khān, against whom Nādir Shāh had to send an expedition about 1734; and Azād Khān, who died in 1885. There is evidence that, in the time of Nādir Shāh, Khārān was still included in

Kirmān; but Nasīr Khān I appears to have brought it under the control of Kalāt, and the country remained under that State until quarrels between Mīr Khudādād Khān and Azād Khān in the middle of the nineteenth century threw the latter into the arms of Afghānistān. In 1884 Sir Robert Sandeman visited Khārān and succeeded in settling the chief points of difference between the chief and Khudādād Khān. Khārān was brought under the political control of the British, and an allowance of Rs. 6,000 per annum was given to the chief. The only Europeans who had previously visited Khārān were Pottinger, who marched through the whole length of the country in 1810, and Macgregor, who crossed the western end in 1877.

The principal objects of archaeological interest are tombs, attributed to the Kiānian Maliks, bearing large brick slabs on which are engraved rough representations of camels, horses, and other animals, the best preserved being at Gwachig in Dehgwar. Inscriptions, presumably Kufic, have been found in Jālwār and Kallag.

The normal population is about 19,000 persons, but it is estimated that 5,500 have recently emigrated. Almost all are nomads living in mat huts and blanket tents. The permanent villages number twenty. The head-quarters of the country are at Shahr-i-Kārez or Khārān Kalāt, which possesses a population of 1,500. Baluchi is the language of the majority, but in the east Brāhui is also spoken. The name usually applied by the people to themselves is Rakhshāni; but this term is strictly applicable only to the groups forming the majority, the remainder being Muhammad Hasnis, and miscellaneous groups such as Kambrānis, Gurgnāris, Chhanāls, Loris, and servile dependants. The dominant class, the Nausher-wānis, consists of nine families. Other Nausherwānis live in Makrān, where their quarrels with the Gichkīs have long been a thorn in the side of the Makrān administration. Camel-breeding and flock-owning are the principal occupations, in addition to agriculture. Felts, rugs in the *dari* stitch, and sacking are made by the women for home use. By religion the people are Sunni Muhammadans. *The people, their tribes and occupations.*

The country is divided into six *niābats*: Khārān with Sarawān, Gwāsh, Shimshān with Salāmbek, Hurmāgai including Jālwār, Māshkel, and Wāshuk with Palantāk. Rāghai and Rakhshān in Makrān also belong to the Khārān chief, and he holds lands in Panjgūr, Mashkai, and elsewhere in the Jhalawān country.

Agriculture.   The greater part of the cultivable area is 'dry' crop, dependent on flood irrigation. Four dams have been constructed in the Baddo river, and one each in the Korakān and the Garruk. The *niābats* of Khārān with Sarawān, Gwāsh, and Wāshuk with Palantāk possess a few irrigated lands. The alluvial soil is fertile when irrigated. The spring harvest consists of wheat with a little barley. In summer *jowār* and melons are grown. Wāshuk and Māshkel contain large date-groves. The system of planting the date-trees is peculiar, the root-suckers being placed in pits, dug to the depth of the moisture-bearing strata, which are kept clear of the wind-blown sand until the suckers have taken root, when the pits are allowed to fill. Camels, sheep, and goats form the live-stock of the country, and are sold in Afghānistān and many parts of Baluchistān. About 100 horses are kept by the chief. Bullocks are few in number. Good salt is obtained from Wād-i-Sultān and Wādiān in the Hāmūn-i-Māshkel.

Commerce and communications.   Since the recent development of Nushki, much of the trade finds its way to that place. Trade is also carried on with Nāl in the Jhalawān country and Panjgūr in Makrān. The exports consist of *ghī* and wool, and the imports of piece-goods, tobacco, and grain, the latter chiefly from the Helmand valley. Sheep and goats are sent to Quetta and Karāchi. Tracks, possessing a moderate supply of water from wells, connect Shahr-i-Kārez with Ladgasht and thence with Panjgūr; with Nāl via Beseima; and with Panjgūr via Wāshuk.

Famine.   Long periods of drought are common, causing the people to migrate. That such migrations were not unknown in former days also is indicated by a *sanad* from Ahmad Shāh Durrāni, which is still extant, permitting the Khārān chief to collect his scattered people from the adjoining countries. In recent years the rainfall has been constantly deficient and much emigration has taken place. The chief always keeps the granaries in his *niābats* full, and when scarcity occurs makes advances in grain without interest, which are recovered at the next harvest. This system is quite exceptional for Baluchistān.

Administration and staff.   In 1884 the chief consented to sit in Kalāt *darbārs* with the Sarawān division of the Brāhuis; but since then he has acquired a position of *quasi*-independence, and is directly controlled by the Political Agent in Kalāt. Each of the *niābats* already mentioned is in charge of a *naib*, whose business is to collect the revenue, pursue raiders and offenders, and report cases after inquiry to the chief or to his agent, known as the *shāh-ghāsi*. Civil cases are decided either by the chief or his agent, or by the *kāzi* at Khāran Kalāt in accordance with Muhammadan law. Order is maintained by a force of about 450 men, armed with swords, matchlocks, and breechloaders. About 170 of these form the garrison of Dehgwar, to prevent raids by the Dāmanis of the Persian border, and 69 are stationed in Rāghai and Rakhshān. In addition, all the tribesmen are liable to military service, when called upon. Those living near Shahr-i-Kārez and all sepoys must always keep ready for emergencies a skin of water, a pair of sandals, and a bag containing about 8 lb. of flour. The chief possesses three muzzle-loading cannon and a mortar.

Besides an allowance of Rs. 6,000 from the Government, the chief's revenue consists of his share of grain in kind; a poll-tax on some households; a goat, sheep, or felt from others; the equivalent of the price of one or two camels from certain groups; fines; unclaimed property; and transit-dues. The aggregate income from local sources fluctuates with the character of the agricultural seasons, but probably amounts to about a lakh of rupees in a good year. The land revenue is levied at the rate of one-fourth to one-tenth of the produce. The chief's own lands are cultivated by his dependants and servants, who receive a share of the produce, generally one-fifth. The largest items of expenditure are incurred on the maintenance of the chief's permanent force, which is estimated to cost about Rs. 2,000 a month, and on the entertainment of guests, the system of Baloch hospitality obliging the chief to keep his house open to all comers.

Revenue.

Makrān (*Makkurān*).—The south-western division of the Kalāt State, Baluchistan, lying between 25° 1' and 27° 21' N. and 61° 39' and 65° 36' E., with an area of about 26,000 square miles. It is bounded on the east by the Jhalawān country and part of Las Bela; on the west by Persia; on the north by the Siāhān range, which separates it from Khāran; and on the south by the sea. The coast-line, which stretches dry and arid from Kalmat to Gwetter Bay, is about 200 miles long. Much of the country consists of mountains, the parallel ranges of which have a general direction east to west, enclosing narrow valleys. The more important are the MAKRĀN COAST, CENTRAL MAKRĀN, and SIĀHĀN ranges. They gradually ascend in height, as they leave the sea, to an elevation of about 7,000 feet. Within them lie the cultivated areas of the country, including Kulānch; Dasht; Nigwar; Kech, also known as Kej, of which Kolwa, Sāmi, Tump, and Mand form part; and Panjgūr with Rakhshān. The CENTRAL MAKRĀN

Boundaries, configuration, and hill and river systems.

hills contain the minor cultivable tracts of Buleda, Bālgattar, Parom, Gichk, and Rāghai. The most important rivers are the DASHT and the RAKSHĀN. They are dry throughout the greater part of the year, but carry heavy floods, and one of their features is the frequent pools from which water is drawn off for purposes of irrigation. Among streams of minor importance may be mentioned the Shādi Kaur, which enters the sea near Pasni, and the Basol, which breaks through the Makrān Coast range. GWĀDAR and PASNI are the seaports, and a little traffic is carried on at Jīwnri. The coast is open and exposed, and owing to the shoaling of the water no large steamers can approach nearer than two miles from the shore.

Geology, botany, and fauna. The only information we possess about the geology of the country is derived from Dr. Blanford's observations[1]. It is known to contain a large development of eocene flysch (Khojak shales), while along the coast the Siwāliks include numerous intercalations of marine strata, known as the Makrān group, containing rich fossil fauna of upper miocene age. The coast appears to coincide with a line of faulting, and the mud-volcanoes, which occur near it, are probably connected with this fracture. The vegetation of the country is similar to that which occurs generally throughout Southern Baluchistān, consisting of an ill-favoured, spiny scrub. Such species as *Capparis aphylla*, *Salvadora oleoides*, *Zizyphus Jujuba*, *Prosopis spicigera*, *Acanthodium spicatum*, *Tamarix articulata*, several kinds of *Acacia*, and many *Astragali* are common. The mangrove grows in the swamps on the coast. Sind ibex and mountain sheep are common in the hills, and ravine-deer along their skirts. An occasional leopard is killed, and wild hog are to be found in places.

Climate, temperature, and rainfall. The climate is marked by three zones of very different character. Along the coast it is uniform and, though hot, not unpleasant. In Kech the winter is healthy and dry, but the heat in summer is intense and in remarkable contrast to the milder atmosphere of the coast. Panjgūr lies in the most temperate zone, with severe cold in winter and moderate heat in summer. The north wind (*gorīch*) is experienced everywhere throughout the year. It is scorching in summer and cutting in winter. During the winter Kech is subject to dense fogs, called *nod*; and, to guard against the damp and the mosquitoes, every native of Makrān possesses a mosquito-curtain. The rainfall is capricious and uncertain, and the country is liable to

[1] *Records of the Geological Survey of India*, vol. v; and *Eastern Persia* (1876).

long periods of drought. Previous to 1904 good rainfall had not been received in Kolwa, Kulānch, and Dasht for five years, and this is said to be no uncommon occurrence. The two periods during which rain is expected are known as *bashshām* and *bahārgāh*. *Bashshām* brings the summer rains, between May 15 and September 15, which generally affect the eastern side of the country. The north and west are more dependent on the winter rains (*bahārgāh*), falling between November and February.

Makrān is generally known as Kech-Makrān, to distinguish History. it from Persian Makrān. Kech-Makrān and Persian Makrān together constitute the Makrānāt, a term occurring in several histories. The etymology of the name is uncertain. By some Makrān is said to be a corruption of *māhi khorān*, 'fish-eaters,' identifiable with the *Ichthyophagi* of Arrian. Lord Curzon considers the name to be Dravidian, and remarks that it appears as *Makara* in the *Brihat Sanhita* of Varāha Mihira in a list of tribes contiguous to India on the west. To the Greeks the country was known as Gedrosia. Lying on the high road from the west to the east, Makrān is the part of Baluchistān round which its most interesting history centres. Legendary stories tell of the marches of Cyrus and Semiramis through its inhospitable wastes, marches which Alexander sought to emulate when he made his famous retreat from India in 325 B.C., so graphically described by Arrian. The *Shāhnāma* relates how Kaikhusrū of Persia took the country from Afrāsiāb of Turān; and the memory of the former, and of his grandfather Kai-Kaus, is preserved in the names of the Khusravi and Kausi *kārez* in Kech. But the suzerainty over Makrān gravitated sometimes to the west, and sometimes to the east. At one time the Sassanian power was in possession; later we hear of its conquest by Rai Chach of Sind. The Arabs, in the seventh century, made themselves masters of the country; but, on the decline of the Caliphate, it disappears from authentic history until Marco Polo mentions it about 1290 as the most westerly part of India under an independent chief. Local tradition relates that of the indigenous races the Rinds, Hots, and Maliks successively held sway in the country after the Arabs; the Maliks were followed by the Buledais, who in their turn were ousted by the Gichkīs from India. In the time of Ahmad Shāh Durrāni, the country was reckoned in the province of Kirmān. Owing to internal dissensions, the Gichkīs fell under the suzerainty of Kalāt in the middle of the eighteenth century; and Mīr Nasīr Khān I acquired the right to half of the revenue of the country, besides extending his conquests westward into Persian Makrān. In 1862 Makrān came into the prominent notice of the British Government in connexion with the construction of the Indo-European Telegraph line, and a British officer was stationed at Gwādar from 1863 to 1871. Meanwhile Persia was extending her power eastward, and in 1879 it was found necessary to depute Colonel Goldsmid to settle the western boundary. Internally matters had gone from bad to worse, owing to the disputes between the Khān of Kalāt and the dominant races, the Gichkīs, Nausherwānis and others, until at length a settlement was effected by Sir Robert Sandeman in 1884. The interference of the British Government has ever since been constantly required, and frequent visits have been paid to the country by European officers supported by escorts. In 1891 Mr. Tate, of the Survey of India, was appointed as the Khān's representative ; but he was withdrawn in 1892, being succeeded by a Hindu Government official as the Khān's *nāzim*. A rising of the Makrānis took place in 1898, when the *nāzim* was temporarily captured, but the rebels shortly afterwards received a severe lesson at the fight of Gokprosh. A Brāhui of good family was thereupon appointed *nāzim*. A disturbance in 1901 led to another small expedition, which captured Nodiz fort. An Assistant Political Agent, who is *ex officio* Commandant of the Makrān Levy Corps, has been posted to Panjgūr since 1904.

The people, their tribes and occupations.

From careful inquiries made in 1903 the population of Makrān was estimated at about 78,000. The permanent villages number 125, the chief of which are Turbat, the headquarters of the administration, Gwādar, Pasni, and Isai. The more important villages are those clustering round the forts, which number fifteen. The population may be divided into five classes : the dominant races ; the middle-class cultivators, generally known as Baloch ; cultivators of irrigated lands, menials, and artisans, called Darzādas, Nakības, and Loris; fishermen, known as Meds and Koras ; and dependants of servile origin. It is distributed into groups, each of which lives independently of the rest; and the democratic tribal system, which is so strongly prevalent in other parts of the Kalāt State, is here non-existent. The dominant races include the Gichkīs, Nausherwānis, Bīzanjaus, and Mīrwāris, the whole of whom probably do not number more than about 500 persons. Their influence is due either to their acquisition of the country by conquest, or to the fact that they represent the ruling power in Kalāt. They are strictly endogamous, and Gichkīs born of Baloch mothers are known as *tolag*, i. e. 'jackal' Gichkīs, and lose much of their social status. The Baloch are the peasant proprietors; the more important are the Hot, Kauhdai, Shehzāda, Kalmati, and Rais. The Darzādas and Nakības are regarded as of aboriginal descent. They are courageous and of fine physique. Of the coast population, the Meds are fishermen and the Koras seamen who make voyages in their vessels to distant countries. Servile dependants abound, and do much of the cultivation and all the household work for men of means. Many of them are Baloch or descendants of Baloch who were captured in the frequent raids which took place in pre-British days. About half of the people are Sunni Muhammadans and the other half Zikris, a curious sect whose alleged incestuous and other immoral practices appear to have been much exaggerated. The language of the country is Baluchi. The majority of the population live by agriculture. Other occupations are flock-owning, seafaring and fishing, weaving, and pottery-making.

Most of the cultivable land consists of 'dry' crop area. Irrigation exists in Kech and Panjgūr, which could probably be improved and developed. Its sources are underground channels (*kārez*), channels cut from pools in rivers (*kaur-jo*), and springs. The *kārez* in use number 127 and the channels cut from rivers 118. The best soil, known as *milk*, consists of a soft white clay. When it contains a mixture of sand it is known as *zawār*. The principal spring crops (*jopāg*) are wheat and barley. Minor crops include beans and pulses. The chief autumn crop (*er-aht*) is *jowār*; rice is cultivated in Kech, Buleda, Panjgūr, and Zāmurān; while Tump, Dasht, and Kulānch produce cotton. The date, however, is the crop *par excellence* of Makrān, and the best are said, even by the Arabs, to surpass those of Basra. The cultivators are well versed in the artificial impregnation of the date-spathes, on which the quality of the produce depends. *Amen*, the date-harvest from July to September, is the pivot round which the thoughts of all the people of Makrān centre, and is a signal for a general influx of all the inhabitants of the surrounding country to Kech and Panjgūr. Horses, camels, cows, donkeys, every beast and every man lives on dates. *Laghati*, or compressed dates, constitutes the staple food of the poor. Those preserved with date-juice in earthen jars, called *humb*, are much relished everywhere. More than 300,000 date-trees are assessed to revenue by the Khān, but the actual number exceeds this

General agricultural conditions and crops.

figure. The Makrāni is an able, though indolent, cultivator, and with the introduction of peace and security agriculture will doubtless develop.

Horses, camels, and sheep. Horse-breeding is not so popular as elsewhere in Baluchistān, and few mares are kept. The breed of cattle is small and generally of brown colour. Makrān donkeys are known for their fleetness. Goods are carried chiefly by camels, which are available everywhere, except along the coast. The commonest sheep in the country are white. Brown and grey sheep, known as *bor* and *kīrg*, are especially valued for their wool, which is made into overcoats (*shāl*). Four-horned sheep are not uncommon in Dasht and Nigwar.

Forests and minerals. No system of forest reservation has yet been introduced. The commonest trees are the tamarisk, which abounds in river-beds, and the acacia. No minerals of economic value have yet been found.

Arts and manufactures. The people comprising the artisan class are generally landholders also. They have no stock in trade and merely supply manufactured articles from the material furnished to them. The weaving industry is moribund, owing to the importation of European cloth. A few coarse cottons are, however, still manufactured. Kerchiefs, used by the women to put over their hair, are made from floss silk obtained from Sarbāz in Persia. Horse-cloths, sword-belts, and shoes are embroidered in silk. The pottery is of the roughest description, consisting of round pitchers and earthen jars.

Commerce. In 1902–3 the imports to the Makrān ports from India were valued at 6½ lakhs and the exports at 7 lakhs. These figures, however, include the trade with the ports of Sonmiāni and Ormāra in Las Bela. No statistics are available regarding the trade which is carried on with places in the Persian Gulf, Arabia, and Africa. The chief centres are Gwādar, where the largest transactions take place; Pasni, Turbat, and Isai. Wholesale trade is carried on entirely by Hindus from Sind and Khojas from Cutch Māndvi. The retail trade is mostly in the hands of Hindus, but a few of the indigenous inhabitants and some Bābis from Kalāt are also engaged in it. The principal exports are raw wool, hides, cotton, matting, dates, salt fish, fish-maws, and shark-fins; and the chief imports are piece-goods and grain, including large quantities of wheat, rice, and *jowār*.

Communications. The communications consist solely of caravan routes, most of which are exceedingly bad, especially those from north to south, which cross the hill ranges at right angles. The main road from Quetta to Bāmpūr in Persia passes through the

Panjgūr valley; another important route between Karāchi, Las Bela, and the west traverses the Kolwa and Kech valleys and eventually also reaches Bāmpūr. Routes from Gwādar and Pasni converge on Turbat northwards. The latter has been recently improved under skilled supervision, and is being continued to Panjgūr through Buleda. Another track from Turbat reaches Panjgūr through Bālgattar. Steamers of the British India Steam Navigation Company carrying the mails call at Pasni and Gwādar on alternate weeks. Both these places have post and telegraph offices.

The production of grain in the country is probably insufficient Famine. for its requirements, but a good date-harvest is enough to meet the needs of the scanty population for the year. In times of scarcity the inhabitants, rapidly dispersing, find a plentiful demand for labour at Karāchi and in Rājputāna.

The administration of the country is conducted, on behalf Adminis- of the Khān of Kalāt, by an officer known as the *nāzim*. He $\frac{\text{tration,}}{\text{staff, and}}$ is assisted by four *naibs*, who are stationed in Tump, Kolwa, levies. Pasni, and Panjgūr. Irregular levies are maintained, number-ing seventy-nine horse and eighty-one foot, and costing about Rs. 32,000 per annum. The infantry hold the important forts of Turbat, Nasīrābād, and Tump in Kech, Bit in Buleda, and Isai in Panjgūr. All persons, including holders of revenue-free grants, are bound to assist the *nāzim* with armed men when occasion requires. For this purpose allowances amount-ing to Rs. 6,000 per annum are granted to certain leading men by the Khān. A telegraph subsidy of Rs. 5,520 is paid by the British Government for the protection of the Indo-European Telegraph line. A Levy Corps of 300 men under two British officers, with its head-quarters in Panjgūr, is being stationed along the western frontier. Its cost, about 1·2 lakhs per annum, is borne by the British Government. Disputes are generally referred to *kāzīs* for decision according to the Muhammadan law. Important awards are confirmed by the Political Agent in Kalāt. Crime is conspicuous by its absence, the number of criminal cases decided in 1900–1 being only sixty-three. The total cost of administration, including the pay of the irregular levies, is about Rs. 80,000 per annum.

It has been stated that Nasīr Khān I obtained from the Land Gichkīs only a right to the collection of half the revenues $\frac{\text{Land}}{\text{revenue.}}$ of the country. In the course of the long series of struggles between the Khāns of Kalāt and the dominant groups which followed, the position gradually changed; and the Khān has now obtained, by confiscation, exchange, &c., the exclusive right to the revenue of some places, while retaining the right to a moiety in others. Elsewhere, the dominant classes hold exclusive rights to collect. The revenue is taken by the appraisement of cereals, the state share being generally one-tenth; by contract; and by a cash assessment on irrigated lands, known as *zarr-e-shāh*, which has now degenerated into a poll-tax of very unequal incidence. A cash assessment is levied on date-trees, and grazing tax is collected at the rate of one sheep in forty or fifty and one seer of *ghī.* Among other receipts are transit-dues, tithes in kind on all fresh fish caught on the coast, and duties on imports and exports. In 1902–3 the total revenue derived from the country by the Khān was Rs. 45,500, to which a grant of Rs. 36,700 was added by him to meet the expenses of administration.

Education, medical, &c.
A little education is imparted by a few ignorant *mullās* and *kāzīs*, generally Darzādas and Afghāns. A Hospital Assistant is attached to the *nāzim*, who affords medical relief in a few cases. The people are very superstitious and attribute almost all diseases to evil spirits, for casting out which special processes are employed. Night-blindness, which is attributed by the people to their diet of fish and dates, is common. Fevers, sore eyes, and ulcers are of constant occurrence. Cholera and small-pox not infrequently visit the country. Vaccination is unknown, but inoculation is popular, the usual fee being four annas. The people thoroughly understand the value of segregation, and careful precautions are taken against the transport of infection by flies.

[Ross: *Memorandum on Makrān*, Bombay, 1867.—*East and West*, vol. iii, No. 31, May, 1904, contains an account of the ancient history of the country by Shams-ul-ulama J. J. Modi.]

Gwādar.—An open roadstead and port in Makrān, Baluchistān, situated in 25° 8′ N. and 62° 19′ E., about 290 miles from Karāchi, with a population of 4,350 persons (1903). The majority are fishermen, Meds. The Portuguese attacked and burnt the town in 1651; and at the end of the next century it was taken by the Khāns of Kalāt and was handed over by Nasīr Khān I to Sultān Said, a brother of the ruler of Maskat, for his maintenance. It has since remained, with about 300 square miles of the adjoining country, in the hands of Maskat, the ruler of which place is represented by an Arab governor, or *wāli*, with an escort of twenty sepoys. The value of the trade, which is carried on by Hindus and by Khojas, locally known as Lotiās, was estimated in 1903 at 5½ lakhs of exports and 2 lakhs of imports. The contract for customs, which are generally levied at 5 per cent. *ad valorem*, was leased for Rs. 40,000 in the same year. Gwādar is a fortnightly port of call of the British India Steam Navigation Company's steamers. On the hill overlooking the town is a stone dam of fine workmanship.

Pasni.—An open roadstead and port in Makrān, Baluchistān, situated in 25° 16′ N. and 63° 28′ E., about 220 miles from Karāchi, on a sandbank connecting the headland of Zarren with the mainland. The inhabitants live in mat huts; the telegraph bungalow and three other structures constitute the only permanent buildings. The population (1904) numbers 1,489, and consists of Meds (1,065) with a few Hindus, Khojas or Lotiās, and Kalmatis. Pasni obtains its importance from its proximity to Turbat, the head-quarters of Makrān, about 70 miles distant. Mail steamers make fortnightly calls at the port, but the open roadstead affords poor anchorage. Improved facilities for landing are now in contemplation. The trade of Pasni is rapidly expanding, and amounted in value to about 4¾ lakhs during the twenty-one months from June, 1903, to February, 1905. The annual customs lease has also risen from Rs. 4,500 in 1899 to Rs. 18,000 in 1905. The only industry is fishing, on which the bulk of the population subsists.

Las Bela.—A small Native State on the southern coast of Baluchistan, lying between 24° 54′ and 26° 39′ N. and 64° 7′ and 67° 29′ E., with an area of 6,441 square miles. It is bounded on the north by the Jhalawān division of the Kalāt State; on the south by the Arabian Sea; on the east by the Kīrthar range, which separates it from Sind; and on the west by the Hālā offshoot of the Pab range. The whole of the eastern part of the State is mountainous; the centre consists of a triangular level plain with its base on the sea; on the west the State has a narrow strip of coast stretching past Ormāra. The hills include the western slopes of the KĪRTHAR mountains as far north as Lak Phusi; the main ridge of the PAB range, with part of the Khude or Khudo and the whole of the Mor offshoot; and on the west the lower slopes of the MAKRĀN COAST range, including Tāloi and Batt. The largest rivers are the PORĀLI and HAB. Minor streams include the Windar, Kharrari, and Phor. The floods of the Windar, Kharrari, and Porāli afford most of the irrigation in the central plain. The Porāli carries a small permanent supply of water at Welpat. The HINGOL is another river which falls into the sea within the State limits. The coast-line extends from the mouth of the Hab river westward

Boundaries, configuration, and hill and river systems.

for about 250 miles, and possesses two roadsteads in SONMIĀNI and Ormāra. Close to the former lies the large backwater known as Miāni Hor. A little to the north of Sonmiāni is the SIRANDA lake.

Geology, botany, and fauna.   The State has never been geologically examined. Alluvial deposits cover the central plain, while the hills consist chiefly of limestone. The vegetation consists of a desolate scrub, represented by such plants as *Boucerosia Aucheriana, Capparis aphylla, Prosopis spicigera, Salvadora oleoides, Acacia farnesiana,* and many *Astragali.* Mangrove swamps occur on the coast. Sind ibex and mountain sheep are numerous in the hills. Ravine-deer are plentiful, and some hyenas, wolves, and wild hog occur. Pangolins are not uncommon. Black and grey partridge afford good sport. Many varieties of fish are caught off the coast.

Climate, temperature, and rainfall.   The climate of the northern parts is extremely hot for eight months of the year. From November to February the air is crisp and cool, causing pneumonia among the ill-clad inhabitants. Along the coast a more moderate and moist climate prevails. Throughout the summer a sea-breeze springs up at midday, to catch which the houses of all the better classes are provided with windsails in the roof. The rainfall is capricious and uncertain, and probably does not exceed the annual average of Karāchi, or about 5 inches.

History and archaeology.   Reference has been made in the article on BALUCHISTĀN to the march of Alexander the Great in 325 B.C. through the southern part of the State. We know that the ruler in the seventh century was a Buddhist priest. The country lay on the route followed by the Arab general, Muhammad bin Kāsim; and Buddhism probably gave place to Islām about this time. The succeeding period is lost in obscurity; but chiefs of the Gūjar, Rūnjhā, Gūngā, and Burfat tribes, which are still to be found in Las Bela, are said to have exercised a semi-independent sway previous to the rise of the Aliāni family of the Jāmot tribe of Kureshi Arabs, to which the present ruling chief, known as the Jām, belongs. The following is the list of the Aliāni Jāms :—

1. Jām Alī Khān I (surnamed Kathūria), 1742-3.
2. Jām Ghulām Shāh, 1765-6.
3. Jām Mīr Khān I, 1776.
4. Jām Alī Khān II, 1818.
5. Jām Mīr Khān II, *circa* 1830.
6. Jām Alī Khān III, 1888.
7. Jām Mīr Kamāl Khān, 1896 (ruling 1906).

The most prominent of these Jāms was Jām Mīr Khān II,

who proved himself a skilful organizer during his long reign. He allied himself with the chiefs of the Jhalawān country in three rebellions against Mīr Khudādād Khān of Kalāt, but in 1869 he was obliged to fly to British territory. In 1877 he was restored to the *masnad*. On his death in 1888 the appointment of a Wazīr, selected by the British Government, was created to assist his successor. The ruling Jām, Mīr Kamāl Khān, did not receive full powers at his accession, but since 1902 they have been increased. The existing relations of the State with the British Government have been detailed in the section on Native States in the article on BALUCHISTĀN.

The shrines of HINGLĀJ and Shāh Bilāwal; the caves at Gondrāni, north of Bela, hewn out of the solid conglomerate rock and possibly of Buddhist origin; and the highly ornamented tombs at Hinidān and other places, affording evidence of a system of superterrene burial, constitute the more important archaeological remains in the State.

Las Bela is divided into seven *niābats*: Welpat, Uthal, Sheh-Liāri, Miāni, Hab, Kanrāch, and Ormāra. It also includes the Levy Tracts along the Hab valley. The permanent villages number 139; the population (1901) is 56,109. BELA is the capital town; SONMIĀNI, Uthal, Liāri, and Ormāra are the only other places of importance. The language is Jadgāli, closely allied to Sindī. Some Baluchi is spoken on the coast. The majority of the inhabitants are Sunni Muhammadans; along the coast are a good many Zikris; a few Khojas and Hindus are engaged in trade. The Gadrās (7,900), who are distinctly negritic in type and generally servile dependants or freedmen, indulge in a kind of fetish worship said to have been brought from Africa. The population is distributed into tribal groups, none of which is, however, numerically strong. The principal are the Jāmot (2,900), Rūnjhā (3,800), and Angāria (2,700). The latter, together with the Sābrā, Gūngā, Burrā, Achrā, Dodā, and Māndrā, are termed Numriā, and are believed to be the aborigines of the country. Landholders and agriculturists compose about half the population; about a quarter are engaged in sheep- and goat-breeding; while the rest of the people are fishermen, traders, labourers, and servile dependants.

The people, their tribes and occupations.

The soil of the country is a fertile, sandy alluvium. Almost the whole of the land depends on flood irrigation, and for this purpose embankments have been constructed in all the principal rivers except the Hab. The area irrigated from

Agriculture.

permanent irrigation is small, and most of it lies in the Welpat *niabāt*. The number of wells is insignificant. They are worked with a leathern bucket and bullocks. The land is in the hands of peasant proprietors. Cash rents are unknown. Tenants, where they exist, receive a share of the grain-heap. Cash wages, except for agricultural labour, which is remunerated in kind, are now coming into vogue. The rates vary from five annas per diem for a common labourer to ten annas for a potter. The staple food-grain of the country is *jowār*, mixed with which *mūng* is grown. These crops constitute the autumn harvest, while the spring harvest consists chiefly of oilseeds.

Cattle, sheep, and goats.   Sheep, goats, and camels are bred in large numbers, especially the two latter. Camels are used for both transport and riding. Horses and ponies are few in number. Bullocks and cows of moderate size are kept for agricultural purposes. Fishing forms an important industry along the coast.

Forests and minerals.   The forests are not systematically 'reserved.' The State derives a small income from those at Malān and Batt, and from the mangroves which grow in the swamps along the coast. In years of good rainfall much excellent forage grass grows on the lower hills and is exported to Karāchi. The minor forest products are gum-arabic, bdellium, and honey. Little is known about the minerals in the State. Marcasite is of frequent occurrence, but not in quantities sufficient to be of commercial value. Limestone is burnt and exported to Karāchi, the State deriving about Rs. 1,500 per annum from it as duty. Salt is obtained from surface excavations at Brār.

Manufactures and commerce.   Rugs of excellent quality are manufactured in the *darī* stitch, and good embroidery is done on cloth and leather with a steel crochet-needle. Trade finds its way to Karāchi by land, and by sea from Gāgu, Sonmiāni, and Ormāra. Caravans proceed to Makrān to exchange grain for dates. The land trade with Sind was valued in 1902–3 at 6·9 lakhs, exports being 5·6 and imports 1·3 lakhs. No separate figures are available for maritime trade. The imports include piece-goods and food-grains, especially rice; and the exports wool, oilseeds, sheep and goats, *ghi*, and fish-maws.

Communications.   The only road is a track, 101 miles long and 12 feet wide, from the Hab river to Bela. Caravan routes connect Sonmiāni with Ormāra; Bela with Makrān; and Bela with Kalāt via Wad. The Indo-European Telegraph line traverses the coast for 226 miles, with an office at Ormāra. The Jām receives a subsidy of Rs. 8,400 per annum for its protection. A daily post, organized by the State, is carried between

Karāchi and Bela, and a bi-weekly service runs between Liāri and Ormāra. Postal expenditure amounts to about Rs. 4,200 annually, and the receipts from stamps to about Rs. 600.

Las Bela is liable to frequent droughts. The longest in Famine. living memory took place between 1897 and 1900, when large numbers of cattle died and a sum of Rs. 5,000 was spent by the State in relief. The poorer classes at such times resort to Karāchi, where a large demand for labour exists.

A description of the system of administration has been Subdivi given in the paragraphs on Native States in the article sions, staff, on BALUCHISTĀN. The suits tried in 1903 aggregated 1,094, and crime. including 267 criminal, 658 civil, and 169 revenue and miscellaneous cases. Cases are seldom referred to *jirgas*. The most common form of crime is cattle-lifting. Special mention may be made of the administration of the Levy Tracts. They formerly belonged to Kalāt; but, in the struggles which occurred during Mīr Khudādād Khān's reign (1857-93), the Chhutta inhabitants developed raiding propensities, directing their attacks against both Sind and Las Bela. A force of Sind Border Police had been organized in 1872 to guard the frontier, but it was not successful; and in 1884 a system of tribal responsibility, under the direction of the Jām, was introduced, the funds being found by the Indian Government. The Las Bela State has since acquired the right to the collection of transit-dues in this tract, from which it receives an income of between Rs. 2,000 and Rs. 3,000 per annum.

The revenue varies from about 1¾ lakhs to 2¼ lakhs, ac-Finance. cording to the character of the agricultural seasons. The expenditure is generally about 2 lakhs. The principal sources of income are land revenue, about Rs. 85,000, and transitdues, which are levied on a complicated system, about Rs. 95,000. Fisheries produce about Rs. 24,500. The expenditure includes the personal allowances of the Jām, about Rs. 40,000; civil establishments, Rs. 50,000; and military, Rs. 45,000. A surplus of Rs. 1,50,000 has been invested in Government securities. Most of the revenue is collected direct by the State officials, but in some cases contracts are given to local traders.

The systematic organization of the land revenue system is of Land recent growth. Up to the time of Jām Mīr Khān II, military revenue service appears to have been the only obligation on the culti- tration. vators. This chief began by assessing land in the possession of traders, and the assessment has since been extended to lands newly brought under cultivation. Revenue-free grants are held chiefly by Saiyids, Shaikhs, Jāmots, Shāhoks, and Numriās. The system of assessment is by appraisement (*tashkhīs*), the state share being fixed by the *tahsīldār* or his representative. The general rate is one-fourth of the produce. Cultivators of crown lands pay one-third.

Military police, army, levies, &c.

A force of military police, consisting of 104 Punjābis, is maintained at Bela under the orders of the Wazīr. They were raised in 1897 and are armed with Snider carbines. The State troops, known as *Fauj Lāsi*, consist of 212 foot, 36 cavalry, and 5 guns. The *naibs* are assisted by sixty-one local levies, known as *faslī* sepoys. Twelve *chaukīdārs* are also maintained. The tribal service of the Levy Tracts consists of five officers, thirty-five footmen, sixteen mounted men, and five clerks, maintained at a total annual cost to Provincial revenues of about Rs. 10,000. The jail at Bela has accommodation for about seventy prisoners. Bela and Uthal each possess a primary school in which 115 boys are under instruction. A dispensary is maintained at Bela at a cost of about Rs. 1,800 per annum. In 1903, 4,750 patients were treated. The commonest diseases are malarial fever, diseases of the eye, and ulcers. A vaccinator is attached to the dispensary, but vaccination is unpopular. Inoculation is practically unknown.

[J. C. Stocks : *New Journal of Botany*, vol. ii, 1850.—A. W. Hughes: *Baluchistān* (1877).]

Bela.—Capital of the Las Bela State, Baluchistān, and the residence of the Jām, situated in 26° 14′ N. and 66° 19′ E. It lies near the apex of the Las Bela plain, 1½ miles from the Porāli river and 116 miles from Karāchi. The population in 1901 numbered 4,183. The majority were State servants, but 356 Hindus were included. The town is not walled and consists of four or five hundred huts. The Jām's residence, a *tahsīl*, treasury, jail, and lines for the military police are the principal buildings. The ancient name of the town was Armāel or Armābel. Sir Robert Sandeman died at Bela in 1892 and was buried on the south of the town. His tomb, of granite and white English marble, is placed beneath a dome erected by the Jām, and is surrounded by a garden. A small establishment is maintained in the town for purposes of conservancy. Cotton cloth and rice constitute the principal imports; oilseeds, *ghī*, and wool the exports. Bela crochetwork is well-known.

Hinglāj.—The best-known place of pilgrimage in Baluchistān, situated in 25° 30′ N. and 65° 31′ E., below the peak of the same name on the banks of the Hingol river in the Las

Bela State. The shrine, which is dedicated to a goddess known as Nāni by Muhammadans and Pārbatī, Kālī, or Māta by Hindus, lies in a verdant basin and consists of a low castellated mud edifice in a natural cavity. A flight of steps leads to a deeper semicircular cleft, through which pilgrims creep on all fours. Bands of pilgrims, each conducted by a leader known as an *agwā*, make the journey by land from Karāchi. Fees are collected at Miāni by a *bhārti* or hereditary examiner, on behalf of the Las Bela officials, from all except devotees and unmarried girls. The proceeds yield about Rs. 600 to the State annually.

Sonmiāni.—Seaport in the Miāni *niābat* of the Las Bela State, Baluchistān, locally known as Miāni, situated in 25° 25′ N. and 66° 36′ E. It is 50 miles from Karāchi by land, and stands on the east shore of the Miāni Hor, a large backwater extending westward in a semicircle, about 28 miles long and 4 miles broad, and navigable as far as Gāgu. Sonmiāni contained a population of 3,166 in 1901, chiefly fishermen (Mohāna), Hindu traders, and a few artisans. Before the rise of Karāchi, Sonmiāni was important as a place through which much of the trade of Central Asia was carried via Kalāt. In 1805 it was taken and burnt by the Joasmi pirates. A British Agent was stationed here in 1840–1. Exports have much decreased, and are at present confined chiefly to salted fish, fish-maws, and mustard-seed.

ERRATA

In table on p. 103, the density of population for the Fort Sandeman *tahsīl* should be 9, not 11.

In table on p. 112, the density of population for the Bārkhān *tahsīl* should be 11, not 5.

On p. 117, l. 12 of **Bārkhān Tahsīl**, for 'inferior' read 'superior.'

On p. 186, l. 5 of **Gwādar**, for '1651' read '1581.'

INDEX

A.

Achaemenian empire, Baluchistān part of, 11.

Administration (general) : the chief officers of, 58–60; administrative divisions, 59, 60; Districts or Political Agencies in charge of Political Agents, 59; subdivisions, 59; *tahsīls* in charge of *tahsīldārs*, 59; circles in charge of *patwāris*, 59; number, districts, and duties of local administrative officers, 59, 60; administration of Kalāt and Las Bela, 60–62; officers of administration have also judicial functions, 63; cost of general administration, 66, 67.

Administration (local) : of Zhob District, 105, 106; Agency Territory in charge of a Political Agent, 105; subdivisions and staff, 105; civil justice and crime, 105, 106; land revenue, 106; of Loralai District, 115; in charge of a Political Agent and Deputy-Commissioner, 115; its subdivisions and *tahsīls*, 115; its civil and criminal courts, 115; land revenue, 115, 116; local boards, 116; civil, criminal, and land revenue, of Quetta-Pishīn District, 124, 125; of Quetta, 125; Agency Territory of Chāgai District, in charge of Political Assistant, 133, 134; of Bolān Pass District, in charge of Political Agent at Kalāt, 137, 138; of Sibi District, 143, 144; courts of justice, 144; of Kalāt State, 60, 61, 156, 162, 168, 174; of Khārān, 177, 178; of Makrān, 185; of Las Bela, 21, 62, 191, 192.

Afghān Wars, the first (1840), 18, 19; the second (1878), 20, 163.

Afghāns (in Baluchistān) (in their own language, Pashtūns), in early history and tradition, 28, 102, 120; on the throne of Delhi, 13, 15, 28; their original seat, 28, 102; language, 26, 27; sex statistics, 25; marriage customs and civil condition, 25, 26; names and distribution of chief tribes, 28; their present numbers, 28; wide and distant migrations, 24, 122; their tribal system, 29; physical and moral characteristics, 29, 30.

Afghānistān, final (1895–6) demarcation of boundary with, 1, 121; trade with, 51, 53.

Age, statistics of, 24, 25; abnormal maximum age-period, due to presence of alien and temporary population, 25.

Agency Surgeon, 58, 66.

Agency Territories, area of, 1, 60, 87; the Districts and *tahsīls* of, 60, 87; administration of, 60; population, 87.

Agent to Governor-General and Chief Commissioner, 58; the head of local administration, his functions, 58, 59; a list of his predecessors, 58; the chief members of his staff, 58; in finance the powers and rules of Indian Local Governments apply to him, 66.

Agriculture (general), 33–41; its recent increase shown by returns of Census and revenue, 33, 37; the small area available, 33; scanty rain and snowfall necessitate irrigation, permanent or temporary, 34; seed-time and harvest in highlands and plains, 34, 57; principal crops, wheat, *jowār*, melons, and dates, 34–36; crops from land manured and irrigated, 36; cropping and fallowing, 37; fruit, great variety of, 37; implements, 37; failure to introduce horse-ploughs, 37; *takāvi* advances by government, usually for irrigation, 37, 38; indebtedness of hill and plain cultivators, 38; rate of interest charged by Hindu *baniā*, 38; cattle, 38; horses, 38, 39; camels, 39; donkeys, 39, 40; sheep and goats, 40; fodder, 40; fairs, 40; cattle-diseases, 40, 41; irrigation, 41, 42; restrictions upon sale and alienation of land, 71.

Agriculture, crops, and harvests (local) : of Zhob District, 103; of Loralai District, 112, 113; of Quetta-Pishīn District, 122, 123; of Chāgai District, 132; of Bolān Pass District, 137; of Sibi District, 141; of Kalāt State, 154; of Sarawān division, 160, 161; of Kachhi division, 166; of Jhalawān, 170; of Khārān, 178; of Makrān, 183, 184; of Las Bela, 189.

Ahmad Shāh, Durrāni (the Afghān), 13; his expeditions to Persia and India, 15; grants title of Ruler of Zhob to head of Jogizais, 102; defeated at Mastung by Nasīr Khān I (1758), 159; subsequent assault of Kalāt, 159, 164; his *sanad* to Khārān chief, 178; his rule of Makrān, 181.

Ahmadzai Khāns of Kalāt, 13–18; their traditional origin, revolt from Mughals, rise, and organization of W. Baluchistān, 13, 14, 153; beginning of authentic history with Mīr Ahmad (1666–7), 14; a list of his successors with dates to Mīr Mahmūd Khān II (1893), the ruling Khān, 14; always subject to a paramount power, 14, 153, Mughal, Afghān,

38; of horses, 39; of donkeys, 40; of camels, 39; of sheep, 40; of goats, 40.

Public Works department, 75–77; its chequered career between civil and military administration, 75; present organization since 1893, 75, 76; Commanding Royal Engineer, Quetta, in charge of all military and civil public works, 76, his duties and staff, 76; military works, roads, waterworks, and buildings, 76, civil works, canals, courts, hospitals, churches, Residencies, Darbār and other Halls for meetings, 76, 77.

Punjab, boundary with, 1; trade with, 53; road to, 56.

Q.

Quetta town (*Kwatah*, Shāl or Shālkot), capital of Baluchistan Agency, and head-quarters of Quetta-Pishīn District, 74, 77, 78, 129, 130; the position and area of the cantonment and of the civil town, 129, 130; number and religions of its population, 129; its ordinary garrison, 129; its average temperatures, 10; permanent occupation of, in 1876, 17; its seizure and siege (1839) in first Afghān War, 18, 77, 129; its evacuation as part of withdrawal from Afghānistān, 19; growth of population since 1891, 24; the base of operations during second Afghān War, 77; prices of commodities at, 44; the railways to, from Bostān and Nushki, 54, 55, 129; connecting roads, 55, 56; artesian wells at, 41; the only municipality in Baluchistan, 74, 130; the constitution and composition of its committee, 74, 126; its income mainly from octroi, 74, 130; its schools, 130; its hospitals, 76, 84, 130; churches, 77, 130; Residency, 77, 130; Darbār Hall and Sandeman Memorial Hall, 77, 130; its drainage and water supply, 77, 130; its strong fortifications and arsenal, 78; the head-quarters of the fourth division of the Western Command, 77; most of the troops of the Province in garrison at Quetta, 77, 78; compulsory vaccination in, 85; its gymkhana ground, 130.

Quetta-Pishīn, a highland District, 119–130; its area, boundaries, configuration, hill and river systems, 119; botany, geology, meteorology, 119, 120; history and archaeology, 120, 121; its 3 towns and 329 villages, 121; the population of the District and *tahsīls*, 121; language, religion, tribes, and occupations, 121, 122; agriculture, 122, 123; cattle, &c., 123; irrigation, 123; minerals, 123; growing trade and industries, 124, 130; communications, 124; administration, civil, criminal, and of land and local revenue, 124–126; army, police, and jails, 126; education, 127; medical, 127; its head-quarters at Quetta town, 129. Bibliography, 127.

Quetta (and Pishīn), the history of the country, 120, 121; their ancient names, 120; the battle ground between Afghāns and Brāhuis, 120; cession of Quetta to Brāhuis (1740) by Nādir Shāh, 120; relations with British in 1839, 1842, 1876, 1879, 120, 121.

Quetta (and Shorarūd), a subdivision and *tahsīl* of Quetta-Pishīn District, 128; held on perpetual lease from Khān of Kalāt, 128; its area, population, villages, and land revenue, 128; its high valley the best cultivated in Baluchistan, 128.

Quinine, sale of, in pice packets, progressing, 85.

R.

Races, 27–28; chief indigenous races: Afghāns, Brāhuis, and Baloch, 27, of Turko-Iranian family, 27, their physical characteristics, 27, geographical distribution and numbers, 27–28; the Lāsis, 27, 28; other and mixed races, 27.

Rai dynasty, conquered Makrān and northwards towards Kandahār, 12; references to, 171, 181.

Railways, 53–55; originally strategical, 53; Sind-Pishīn Railway a result of second Afghān War, 53, its course, length, and construction, 54–55, the Chappar rift and span bridge, the Khojak tunnel, and Mud Gorge, its branch to Quetta, 54–55; the Bolān Pass extension, 53, its heavy tunnelling and steep gradients, 55; Quetta-Nushki line completed in 1905, 54, 55; their enormous cost, 54, 55; their civilizing and unifying effects, 53; of Quetta-Pishīn District, 124; of Chāgai District, 133; of Bolān Pass District, 137; of Sibi District, 143; of Kalāt State, 155; of Sarawān division, 161.

Rainfall: irregular and scanty owing to absence of monsoon, 10, a table of average rainfall at principal places, 11; uncertainty of protective irrigation dependent upon it, 58; of Zhob District, 101; of Loralai District, 110; of Quetta-Pishīn District, 120; of Chāgai District, 131; of Bolān Pass, 136; of Sibi District, 139, 147; of Kalāt State, 152; of Sarawān division, 159; of Kachhi division, 165; of Jhalawān, 171; of Khārān, 176; of Makrān, 180, 181; of Las Bela, 188.

Rakhshān river, 258 miles long, 3, 95, 99, 100; joins the Mashkel from Persia and runs to Hāmūn-i-Māshkel, 100.

Oxford : Printed at the Clarendon Press by HORACE HART, M.A.

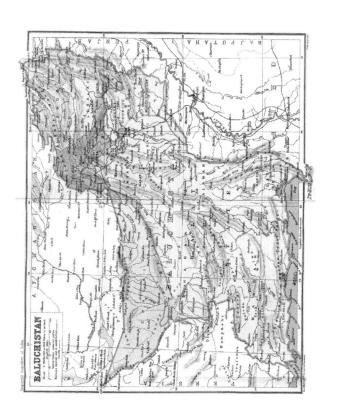

BALUCHISTAN